Andreas Crivellin

Non-minimal flavor-violation in the MSSM

Andreas Crivellin

Non-minimal flavor-violation in the MSSM

Südwestdeutscher Verlag für Hochschulschriften

Imprint
Any brand names and product names mentioned in this book are subject to trademark, brand or patent protection and are trademarks or registered trademarks of their respective holders. The use of brand names, product names, common names, trade names, product descriptions etc. even without a particular marking in this work is in no way to be construed to mean that such names may be regarded as unrestricted in respect of trademark and brand protection legislation and could thus be used by anyone.

Publisher:
Südwestdeutscher Verlag für Hochschulschriften
is a trademark of
Dodo Books Indian Ocean Ltd., member of the OmniScriptum S.R.L Publishing group
str. A.Russo 15, of. 61, Chisinau-2068, Republic of Moldova Europe
Printed at: see last page
ISBN: 978-3-8381-2035-5

Zugl. / Approved by: Karlsruhe, Universität, Dissertation, 2010

Copyright © Andreas Crivellin
Copyright © 2010 Dodo Books Indian Ocean Ltd., member of the OmniScriptum S.R.L Publishing group

ABSTRACT

The MSSM possesses many new sources of flavor violation. In addition to the minimally flavor-violating interactions involving the CKM matrix there are terms which have a priori a generic flavor structure (including possible complex phases) stemming from the supersymmetry-breaking sector.
Especially, the trilinear A-terms can have an important effect on flavor-violating observables because they are both chirality and flavor violating. In the presence of generic sources of flavor-violation processes mediated by squarks and gluinos are of special interest because they involve the strong coupling constant. This would lead to dangerously large effects, which would be in contradiction with experiment if the flavor structure were arbitrary. The difficulty to explain why the soft-supersymmetry breaking terms are approximately universal, or aligned to the quark sector, is known as the "SUSY flavor problem".
An important effect that can amplify corrections in the MSSM is chiral enhancement. Chiral enhancement is always related to self-energies because a large chirality-changing parameter in a SUSY correction occurs instead of a small fermion mass. A parametric enhancement by a factor of $A_{ii}^f/(M_{\mathrm{SUSY}} Y^{f_i})$ or $\tan\beta$ can then (at least partly) compensate for the suppression factor of $\alpha_s/(4\pi)$.
The results of this thesis include the following[1]:

- The finite renormalization of fermion masses and mixing matrices by one-particle irreducible self-energies is presented. We apply this formalism to the chirally enhanced pieces of the MSSM self-energies taking also into account the important NLO corrections to the fermion masses and mixing angles.

- Applying 't Hooft's naturalness argument strong constraints on the soft-supersymmetry-breaking parameters $\delta_{ii}^{f\,LR}$ and $\delta_{ij}^{q\,LR}$ ($j > i$) are obtained by demanding that the SUSY corrections to fermion masses and CKM angles should not exceed the measured values. The NLO effects also allow us to constrain the combinations $\delta_{13}^{f\,LR}\delta_{31}^{f\,LR}$

---
[1]The results of this thesis have already been published for the most part in Ref. [1–6]

and $\delta_{23}^{d\ LR}\delta_{32}^{f\ LR}$. This is especially important with respect to $\delta_{13,23}^{u\ RL}$ which are unconstrained from FCNC processes.

- Chirally enhanced corrections to FCNC processes are computed. We show that all such (N)NLO contributions can easily be included into the LO analysis by renormalizing the squark-quark-gluino vertex. It is found that these corrections are important in the presence of chirality-violating terms if the squark masses are non-degenerate. Taking into account these corrections, new bounds on the flavor-violating mass-insertions $\delta_{12,13,23}^{d\ AB}$ are calculated.

- A model in which the light fermion masses are generated radiatively via sfermion-gaugino loops using the trilinear A-terms is proposed. We show that this model does not only possess a higher flavor symmetry in the Yukawa sector but is also capable to solve the SUSY CP as well as the SUSY flavor problem. The phenomenological consequences of this model are worked out using the improved FCNC analysis which includes the chirally enhanced corrections. It is found that the dominant effects occur in $b \to s(d)\gamma$ and in Kaon mixing. If the model is extended to the lepton sector, the anomalous magnetic moment of the muon requires the smuon mass to be approximately between 1 TeV and 3 TeV.

- The constraints on the mass-splitting of the first two generations of left-handed squarks obtained from ΔM_K, ϵ_K and $D-\overline{D}$ mixing is studied. The different contributions from gluino, neutralino and chargino diagrams are examined in detail, concluding that it is not justified to neglect electroweak gaugino diagrams if the squark mass matrices contain flavor non-diagonal LL elements. We find that the constraints on the mass-splitting are very strong for light gluino masses. However, if the gluino is heavier than the squarks the constraints on the mass-splitting are much weaker. There are even large regions in parameter space where the different new physics contributions cancel each other, leaving the mass-splitting nearly unconstrained.

- We show that a sizable right-handed W coupling can be generated within the MSSM by using the unconstrained elements $\delta_{13,23}^{u\ RL}$. This anomalous coupling effects the extraction of V_{ub} and V_{cb} from inclusive and exclusive decays and is capable to solve the apparent discrepancies between the different determinations of these CKM elements.

Contents

1	**Introduction**	**1**
2	**The MSSM**	**6**
	2.1 Construction of the MSSM	8
	2.2 The mass spectrum of the MSSM	10
	2.2.1 Sfermions	11
	2.2.2 Charginos	13
	2.2.3 Neutralinos	13
3	**Finite renormalization of fermion masses and mixing matrices**	**15**
	3.1 General formalism	15
	3.2 Self-energies in the MSSM	17
	3.3 Mass and wave function renormalization in the MSSM	19
	3.4 Renormalization of the CKM matrix	23
	3.4.1 Super-CKM basis beyond tree-level	24
	3.4.2 The CKM matrix in charged-Higgs and chargino couplings	27
4	**Naturalness constraints**	**30**
	4.1 Constraints on flavor-diagonal mass insertions at one loop	31
	4.2 Constraints on flavor-off-diagonal mass insertions from CKM and PMNS renormalization	31
	4.2.1 CKM matrix	31
	4.2.2 Down-sector	32
	4.2.3 Up-sector	34
	4.2.4 Comparison with previous bounds	34
	4.2.5 Threshold corrections to PMNS matrix	35
	4.3 Constraints from two-loop corrections to fermion masses	37

5 Chirally enhanced corrections to FCNC processes — 45
5.1 One-loop renormalization of the quark-squark-gluino vertex — 46
5.2 $B \to X_s \gamma$ — 50
5.3 $\Delta F = 2$ processes — 53
  5.3.1 $B - \bar{B}$ mixing — 55
  5.3.2 $K - \bar{K}$ mixing — 57

6 Radiative generation of light fermion masses — 62
6.1 Description of the model — 62
6.2 Phenomenological consequences in the quark sector — 64
  6.2.1 CKM generation in the down-sector — 64
  6.2.2 CKM generation in the up-sector — 65
6.3 Phenomenological consequences in the lepton sector — 68

7 LL Mass Insertions and constraints from K and D mixing — 72
7.1 Meson mixing between the first two generations — 73
7.2 Constraints on the mass splitting from Kaon mixing and D mixing. — 75

8 Right-handed W coupling and its effects on V_{ub} and V_{cb} — 81
8.1 Right-handed W couplings — 82
  8.1.1 Determination of V_{ub}^L and V_{cb}^L — 83
8.2 SQCD corrections to the quark-quark-W vertex — 84

9 Conclusions — 88

10 Appendix — 92
10.1 Feynman rules — 92
10.2 Higgs vertex corrections — 93
10.3 Loop integrals — 93

LIST OF FIGURES

2.1 a) Higgs self-energy which a fermion.
b) Higgs self-energy which a scalar particle. 7

2.2 Insertion of a flavor-changing mass term into a sfermion line. 13

3.1 Flavour-valued wave-function renormalization. 16

3.2 One-particle irreducible two-loop self-energy constructed out of two one-loop self-energies. 21

3.3 Tree–level coupling with Y_{ij}^d and FCNC loop corrections with A_{fi}^d and $\Delta_{fi}^{\tilde{d}LL,RR}$ in the mass insertion approximation for $M_{\text{SUSY}} \gg v$. Replacing the Higgs fields by their vevs gives the contributions to the down–type quark mass matrix. 22

3.4 Genuine SQCD vertex correction 23

3.5 One-loop corrections to the CKM matrix from the down and up sectors. . 24

4.1 Constraints on the diagonal mass insertions $\delta_{11,22}^{\ell,u,dLR}$ obtained by applying 't Hooft's naturalness criterion. 38

4.2 Constraints on $|\delta_{12}^{dLR}|$ from $|V_{us}|$ (or $|V_{cd}|$) as a function of the squark mass. 39

4.3 Constraints on $|\delta_{12}^{dLR}|$ from $|V_{us}|$ (or $|V_{cd}|$) as a function of the gluino mass. 39

4.4 Constraints on $|\delta_{23}^{dLR}|$ from $|V_{cb}|$ (or $|V_{ts}|$) as a function of the squark mass. 40

4.5 Constraints on $|\delta_{13}^{dLR}|$ from $|V_{ub}|$ as a function of the squark mass. 40

4.6 Constraints on $|\delta_{13}^{dLL}|$ from $|V_{ub}|$ as a function of the gluino mass for different values of $\mu \tan\beta/(1+\Delta_b)$. 41

4.7 Numerically important two-loop correction to V_{td}. The analogous diagram also exists in the up sector. 41

4.8 Constraints on δ_{12}^{uLR}, δ_{23}^{uLR} and δ_{13}^{uLR} as a function of the squark mass for different ratios of $m_{\tilde{g}}/m_{\tilde{q}}$. 42

LIST OF FIGURES

4.9 $|\Delta U_{e3}|/U_{e3}$ in %. Top: as a function of δ_{13}^{lLR} for $M_{\text{SUSY}} = 1000$ GeV and different values of θ_{13}. Bottom: as a function of θ_{13} for $M_{SUSY} = 1000$ GeV and different values of δ_{LR}^{l13}. 43

4.10 Excluded regions in the $(\theta_{13}, \delta_{13}^{lLR})$-plane for different values on M_{SUSY}. . . 43

4.11 Regions compatible with the results of the two-loop contribution to the electron, up, down and strange mass for different values of M_{SUSY}. 44

5.1   Examples of chirally enhanced two-loop and three-loop diagrams contributing to B_s mixing which can compete with (or even dominate over) the one-loop diagrams. 46

5.2   Br(B $\to$ X$_s\gamma$) as a function of δ_{23}^{dLL} and δ_{23}^{dRR}, respectively, for $m_{\tilde{g}} = 750$ GeV, $m_{\tilde{q}1,2} = 2m_{\tilde{q}3} = 1000$ GeV, $\tan\beta = 50$ and different values of $\mu/(1+\Delta_b)$. 51

5.3   Br(B $\to$ X$_s\gamma$) as a function of δ_{23}^{dLR} and δ_{23}^{dRL}, respectively, for $m_{\tilde{g}} = 750$ GeV, $m_{\tilde{q}1,2} = 2m_{\tilde{q}3} = 1000$ GeV, $\tan\beta = 50$. Yellow(lightest): experimentally allowed range for Br(B $\to$ X$_s\gamma$) and different values of $\mu/(1+\Delta_b)$. 53

5.4 $B \to X_s\gamma$: allowed regions in the $\delta_{23}^{dLR,RL} - \frac{\mu}{1+\Delta_b}$ and $\delta_{23}^{dLL,RR} - \frac{\mu}{1+\Delta_b}$ planes for $m_{\tilde{q}1,2} = 2m_{\tilde{q}3} = 1000$ GeV, $\tan\beta = 50$ and different values of $m_{\tilde{g}}$. 58

5.5   Left: Contour plot of $\Delta M_{q,\text{ren}}/\Delta M_{q,\text{LO}}$ as a function of $m_{\tilde{q}1,\tilde{q}2}$ and $m_{\tilde{q}3}$ with $m_{\tilde{g}} = 1000$ GeV. Right: $\Delta M_{q,\text{ren}}/\Delta M_{B\text{LO}}$ as a function of $m_{\tilde{q}1,\tilde{q}2} = 2m_{\tilde{q}3}$ and $m_{\tilde{g}}$. 59

5.6   Allowed range for NP-contributions to $B_q - \bar{B}_q$ mixing, $q = d, s$, in the $\phi_{B_q} - C_{B_q}$ plane. 59

5.7   Allowed regions in the complex $\delta_{13,23}^{dLR}$-plane from B_s-mixing (left plot) and B_d-mixing (right plot) with $m_{\tilde{q}1,2} = 2m_{\tilde{q}3} = 1000$ GeV for different values of $m_{\tilde{g}}$ 60

5.8   Left: Contour plot of $\Delta M_{K\text{ren}}/\Delta M_{K\text{LO}}$ as a function of $m_{\tilde{q}1}$ and $m_{\tilde{q}2}$ with $m_{\tilde{g}} = 1000$ GeV. Right: $\Delta M_{K\text{ren}}/\Delta M_{K\text{LO}}$ as a function of $m_{\tilde{q}2} = 0.98 m_{\tilde{q}1}$ and $m_{\tilde{g}}$. 60

5.9   Left: Allowed range for NP contributions to K mixing. Right: Allowed region in the complex δ_{12}^{dLR}-plane with $m_{\tilde{g}} = 750$ GeV, $m_{\tilde{q}1} = 1000$ GeV and different values of $m_{\tilde{q}2}$. 61

LIST OF FIGURES

6.1 Allowed regions in the $m_{\tilde{g}} - m_{\tilde{q}}$ plane.
Left: Constraints from $b \to s\gamma$ for different values of $m_b\mu \tan\beta/(1 + \Delta_b)$ assuming that the CKM matrix is generated in the down sector.
Right: Constraints from Kaon mixing for different values of M_2 assuming that the CKM matrix is generated in the up sector. 66

6.2 Predicted branching ratio for the rare Kaon decay $K_L \to \pi\nu\bar{\nu}$ assuming that the CKM matrix is generated in the up-sector for $m_{\tilde{q}} = m_{\tilde{g}}$. 68

6.3 Predicted branching ratio for the rare Kaon decay $K^+ \to \pi^+\nu\bar{\nu}$ assuming that the CKM matrix is generated in the up-sector for $m_{\tilde{q}} = m_{\tilde{g}}$. 69

6.4 Left: Allowed region in the M_1-$m_{\tilde{\mu}}$ plane assuming that the muon Yukawa coupling is generated radiatively by $v_d A_{22}^l$.
Right: Allowed region in the M_1-$m_{\tilde{e}}$ plane assuming that the electron Yukawa coupling is generated radiatively by $v_d A_{11}^l$. 71

7.1 Size of the real part of Wilson coefficients C_{1SUSY}, $C_1^{\tilde{g}\tilde{\chi}^0}$, $C_1^{\tilde{\chi}^+\tilde{\chi}^+}$, $C_1^{\tilde{\chi}^0\tilde{\chi}^0}$ and $C_1^{\tilde{g}\tilde{g}}$ contributing to $D - \bar{D}$ or $K - \bar{K}$ mixing. 79

7.2 Allowed regions in the $m_{\tilde{q}_1} - m_{\tilde{g}}$ plane with $m_{\tilde{q}_{2,3}} = 500, 1000$ GeV for different values of M_2. The maximally allowed mass splitting assuming an intermediate alignment of $m_{\tilde{q}}^2$ with $Y_u^\dagger Y_u$ and $Y_d^\dagger Y_d$, the allowed range assuming an diagonal up (down) squark mass matrix and the minimal allowed region for the mass splitting between the left-handed squarks are shown. 80

8.1 $|V_{ub}^L|$ as a function of Re $\left(V_{ub}^R/V_{ub}^L\right)$ extracted from inclusive decays, $B \to \pi l\nu$ and $B \to \tau\nu$. 83

8.2 $|V_{cb}^L|$ as a function of Re $\left(V_{cb}^R/V_{cb}^L\right)$ extracted from inclusive and exclusive processes. 85

8.3 Feynman diagram which induces the effective right-handed W coupling of quarks. 86

8.4 Left: Relative strength of the induced right-handed coupling $|V_{ub}^R|$ with respect to $|V_{ub}^L|$ for $M_{\text{SUSY}} = 1$ TeV. $|V_{ub}^L|$ is determined from CKM unitarity.
Right: Same as the left figure for V_{cb}^R. 87

1. Introduction

The Standard Model (SM) of particle physics (supplemented with neutrino masses) provides a remarkably successful description of the presently known phenomena. Its predictions are compatible with experiments ranging from laboratory-size low-energy precision experiments up to the largest terascale particle colliders. Electroweak precision data obtained by LEP have established the SM mechanism of spontaneous symmetry breaking (except for the still lacking discovery of the Higgs particle) and direct searches for new particles at the Tevatron experiments CDF and D$\emptyset$ at Fermilab gave no hint for physics beyond the SM. Furthermore, the B-factories BABAR and BELLE have confirmed the Cabibbo-Kobayashi-Maskawa (CKM) mechanism of flavor violation with very high accuracy. Especially, even though Flavour-Changing Neutral Current (FCNC) processes are very sensitive to new physics (NP) (since the SM contribution is strongly suppressed) all results are in reasonable agreement with the SM.

However, it still seems clear that the SM is not the ultimate "theory of everything" since it does not incorporate gravity. Therefore, we already know at least one scale of new physics (NP): The reduced Planck scale $M_P = \sqrt{8\pi G_{\text{Newton}}} = 2.4 \times 10^{18}\text{GeV}$, where gravitational effects become important and classical General Relativity (GR) is no longer valid. Furthermore, there is compelling evidence for a Grand Unified Theory (GUT) in which the three SM gauge couplings unify at some high scale $M_{\text{GUT}} \approx 10^{14}\text{GeV} - 10^{16}\text{GeV}$. In the simplest GUT, SU(5), all SM particles fit nicely into its representations and since SU(5) is a simple group this can explain the quantization of electric charge. In a more complicated GUT, SO(10), also the right-handed neutrinos find their natural places.

The fact that the scale of NP, M_{GUT} or M_P is many orders of magnitude above the electroweak scale raises the infamous "hierarchy problem". However, it should be kept in mind that this is not a problem of the SM itself but rather an intricate sensitivity of the Higgs potential to radiative corrections associated with heavy new particles present in extensions of the SM. The origin of this extreme sensitivity to quantum loop corrections is the fact that the Higgs is a scalar particle whose renormalization is plagued by quadratic divergences.

The best studied (and maybe also the best motivated) solution to the hierarchy problem is supersymmetry (SUSY). It introduces superpartners to all SM particles which have the same couplings and quantum numbers but which differ in spin by 1/2. Since a fermionic loop has the opposite sign as a bosonic one, the contributions to the Higgs mass cancel exactly in the limit of unbroken SUSY in which the particles in a supermultiplet have the same masses. However, we know that SUSY cannot be exact but rather must be broken since no superpartners have been detected (yet). Nevertheless, the renormalization property of the Higgs mass is significantly better since now the divergence depends on the ratio $M_{\text{SUSY}}/M_{\text{SM}}$. However, in order to provide a convincing solution for the hierarchy problem one should expect the SUSY scale to be of the order of one TeV.

Since we can at best speculate about the mechanism that breaks SUSY, explicit soft-supersymmetry-breaking terms are introduced which parameterize our ignorance. Soft means in this context that the terms are chosen in such a way that the non-renormalization theorem, which guarantees the stability of the Higgs potential, is still valid. Yet, the effect of these soft SUSY breaking terms is not only to give (additional) masses to the SUSY particles. In fact, they are also potential new sources of CP and flavor violation since there is a priori no constraint which requires them to be flavor diagonal and real in the same basis as the Yukawa couplings. These new flavor-changing (and CP violating) interactions involve the strong coupling constant (in the quark sector) and since we expect the SUSY scale to be of the order of 1 TeV a fully unconstrained version of the MSSM would be disastrous for the FCNC transitions. Therefore, SUSY-breaking scenarios with alignment or flavor universality were invented. However, this condition is usually imposed at some high scale and is not renormalization group invariant. This means that at the lower (SUSY) scale off-diagonal elements are induced which are proportional to Yukawa matrices.

Anyway, since we expect the SUSY scale to be of the order of 1 TeV, given the strong suppression of FCNC processes it is very interesting (and important for model building) to calculate the resulting bounds on the flavor (and chirality) violating SUSY-breaking parameters [7–27]. The common parameterization for these flavor (and chirality) changing quantities is motivated by the mass insertion approximation and is therefore called the mass insertion parameterization [28]. Even though we will not use this approximation in our analysis it is very helpful for a qualitative understanding of the effects and we will also use the mass insertion parameterization for the off-diagonal elements of the sfermion mass matrices. The mass insertion parameter $\delta_{ij}^{f\,AB}$ is usually defined to be the off-diagonal

1. Introduction

element of the sfermion mass matrix divided by the geometric average of the corresponding diagonal elements (see chapter 2 for details):

$$\delta_{ij}^{f\ AB} = \frac{\Delta_{ij}^{f\ AB}}{M_{i\ A}^{\tilde{f}} M_{f\ B}^{\tilde{f}}} \quad (1.1)$$

Here i and j are flavor indices while A and B denote the chiralities L and R. While all flavor-changing elements in the down sector can be constrained, the elements of the up sector which involve the third generation are rather unconstrained.

If the mass insertion parameter $\delta_{ij}^{f\ AB}$ is chirality changing ($A \neq B$) another interesting and important effect occurs: chiral enhancement. In the flavor-conserving case the non-decoupling chirality-changing part of a SQCD self-energy can be of order one compared to the corresponding quark mass because of a parametric enhancement by a factor of $\tan\beta$ or $A_{ii}^f/(M_{\text{SUSY}} Y^{f_i})$. Also in the flavor-changing case order one effects are possible because one has to compare the self-energy to the corresponding quark mass times the CKM element. Since they can be of order one, the chirally enhanced corrections lead to formally (N)NLO diagrams which can be of the same order as the leading order process. Therefore, these corrections have to be included into the calculation.

Even though the chirality-changing part of the self-energy can be of order one it cannot be distinguished from a mass term in the decoupling limit. However, it corrects the relation between the physical mass and the Yukawa coupling (or between the measured CKM element and the one in the Lagrangian). According to 't Hooft, a small quantity is natural if a symmetry is gained if it is set to zero. Large accidental cancellations, not enforced by symmetry, are unnatural and therefore should not occur in a valid theory. This argument enforces that the SUSY corrections to the fermion masses and to the CKM matrix should not exceed the experimentally measured values. Out of this requirement strong bounds on the mass insertions $\delta_{ij}^{f\ AB}$ can be obtained.

Furthermore, in the case when the SUSY corrections are as large as the measured values one recovers a very interesting scenario: radiative mass generation. Since the A-terms generate the Yukawa couplings via loops they are necessarily aligned to them. In this way the SUSY flavor and the SUSY CP problem can be solved. However, the top quark and the tau lepton are too heavy to be generated radiatively and because of the successful bottom-tau Yukawa coupling unification (and in order to be consistent) it is sensible to keep the bottom Yukawa as well. Therefore, in the quark sector a misalignment between the Yukawa matrices, diag$(0, 0, Y^{q_3})$, and the A-terms, which generate the light quark masses and the

CKM matrix, occurs. This leads to observable deviations from MFV if the third generation is involved. In the lepton sector this model affects the anomalous magnetic moment of the muon (and to less respect also the electron) which receives corrections due to the chirality changing nature of A_{22}^ℓ.

Since flavor-changing self-energies evaluated at zero external momentum can be of order one they also lead to (N)NLO contributions to FCNC processes which can be of the same order as the LO processes. This occurs since we can treat diagrams with FC self-energies in the external legs as one-particle irreducible [29]. Therefore, a inclusion of these effects is important and, as we will show later, can be easily achieved by including the self-energies into an effective squark-quark-gluino vertex.[1] It will turn out that these effects drop out for degenerate squark masses but can be dominant for unequal masses, especially if transitions between the first two generations are involved.

A mass splitting is especially interesting for the left-handed squarks. Already in the early stages of MSSM analyses it was immediately noted, that a super GIM mechanism is needed in order to satisfy the bounds from flavor changing neutral currents (FCNCs) [30]. Therefore, the mass matrix of the left-handed squarks should be (at least approximately) proportional to the unit matrix, since otherwise flavor off-diagonal entries arise inevitably either in the up or in the down sector due to the SU(2) relation between the left-handed squark mass terms. The idea that non-degenerate squarks can still satisfy the FCNC constraints (K and D mixing) was first discussed in Ref. [31] (an updated analysis can be found in Ref. [32]) in the context of abelian flavor symmetries [33, 34]. In all MSSM analyses the main focus has been on the gluino contributions, while the chargino and neutralino contributions were usually neglected claiming that they are suppressed by a factor of g_2^4/g_s^4 [9–11, 25, 31, 35, 36]. However, it is no longer a good approximation to consider only the gluino contributions in the presence of off-diagonal elements in the LL block of the squark mass matrices because the winos couple to left-handed squarks with g_2. In addition, the gluino contributions suffer from cancellations between the crossed and uncrossed box-diagrams, especially if the gluino is heavier than the squarks. Therefore, the neutralino and chargino contributions can even be dominant if M_2 is light and the gluino is heavier than the squarks. This situation can occur in GUT-motivated scenarios in which the relation $M_2 \approx m_{\tilde{g}} \alpha_2/\alpha_s$ holds. Therefore, we want to update the evaluation of the constraints from K and D mixing with focus on the mass splitting between the first two squark generations taking into

---

[1]Similar effects occur in the lepton sector, however they are less important since they don't involve the strong coupling constant.

account the weak contributions as well.

There is another interesting feature of flavor-changing A-terms: They can generate a sizable right-handed W coupling via loops. We find that the right-handed W-coupling can only be sizable with respect to $b \to u$ and $b \to c$ transitions. Furthermore, the decoupling corrections to the left-handed coupling are suppressed due to cancellations between the genuine vertex correction and the self-energy diagrams and cannot be sizable. In opposite to left-right symmetric models [37] the corresponding dimension six operator does not lead to a right-handed coupling to neutrinos. Therefore, the constraints from neutrino masses on a right-handed W coupling do not apply in this case. A generic analysis of such higher-dimensional right-handed couplings has been performed in Ref. [38] aiming at a better understanding of $K \to \pi\mu\nu$ data. The general effect of left- and right-handed anomalous couplings of the W to charm was studies in Ref. [39] and the coupling of the W to up in [40]. We will investigate the effect of a right-handed W-coupling on the extraction of $|V_{ub}|$ ($|V_{cb}|$) and show that current tensions between SM and data can be removed (alleviated). This work is composed as follows: Chapter 2 gives a short introduction to the MSSM with special emphasis on the new sources of flavor violation. Chapter 3 discusses the finite renormalization induced by SQCD self-energies beyond leading order. A numerical evaluation of the fine tuning constraint obtained by applying 't Hooft's naturalness argument is given in chapter 4. Chapter 5 discusses the chirally enhanced corrections to FCNC processes in the presence of generic flavor violation. We use this improved analysis in the study of the phenomenological consequences of a model with radiative mass generation in chapter 6. The importance of the electroweak contributions to $\Delta F = 2$ processes in the presence of LL mass insertions is discussed in chapter 7 with focus on the mass splitting between the first two generations of left-handed squarks. In chapter 8 we show that the MSSM can generate a sizable right-handed W coupling. The effect of such a coupling on the determination of V_{ub} and V_{cb} from inclusive and exclusive decays is discussed in the effective field theory approach. We conclude in chapter 9 and an appendix summarizes the loop functions, Feynman rules etc. needed for the computation.

2. The MSSM

As already noted in the introduction, the SM is for sure not the ultimate "theory of everything". Even the force which is most familiar to us from everyday life, gravity, is not included and quantum gravitational effects become important near the Planck scale. Therefore, we already know at least one scale of new physics: $M_P = \sqrt{8\pi G_{\text{Newton}}} = 2.4 \times 10^{18}\,\text{GeV}$. Furthermore, there is compelling evidence for a grand unified theory:

- The SM fermions fit nicely into the $\bar{5}$ and 10 representation of SU(5) and the gauge bosons can be naturally embedded into the 24 dimensional adjoint representation.

- Taking into account the running of the three SM gauge couplings they (nearly) meet at an energy scale of $10^{14}\,\text{GeV}$.

- The hypercharge quantum numbers of the SM fermions look random. However, combining hypercharge Y with the baryon number B and lepton number L into $Y - (B - L)/2$ one recognizes the pattern of a right-handed isospin. This hints towards a left-right symmetric model which can be embedded into SO(10) GUTs. Furthermore, in SO(10) GUTs the right-handed neutrino is naturally incorporated as a SU(5) singlet.

However, any scale of new physics (M_{Planck}, M_{GUT}) which is far above the scale of electroweak symmetry breaking of the standard model causes the infamous hierarchy problem [41–44]: Let's illustrate this by a simple example. Consider the generic coupling term of a fermion to the Higgs field in the Lagrangian $-\lambda_f \bar{f} f H$. The resulting Higgs self-energy is shown in Fig. 2.1a. If we use a ultraviolet Λ_{UV} cutoff for the regularization procedure we get the following correction to the Higgs mass:

$$m_H^2 \to m_H^2 - \frac{|\lambda_f|^2}{8\pi^2}\Lambda_{UV}^2 + \log\text{div.} + \text{finite} \qquad (2.1)$$

As we see from (2.1), this correction is especially disturbing for the top quark due to its large Yukawa coupling. However, this problem also arises for any new heavy particle (for

2. The MSSM

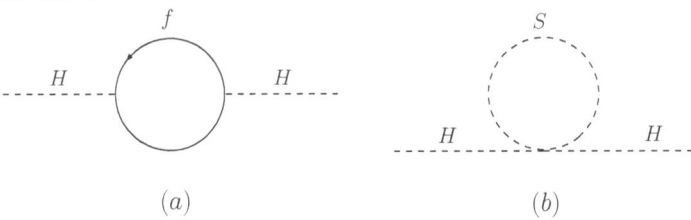

Figure 2.1: a) Higgs self-energy which a fermion leads to a quadratic divergence. b) Higgs self-energy which a scalar particle which also leads to a quadratic divergence but with opposite sign than the one resulting from Fig. a).

example if we have some heavy gauge boson of a GUT). Consider now the coupling of a new complex scalar with the Lagragian term $-\lambda_s |H|^2 |S|^2$. Calculating the corresponding Higgs self-energy (Fig. 2.1b) the correction to the Higgs mass squared reads:

$$m_H^2 \rightarrow m_H^2 + \frac{\lambda_s}{16\pi^2}\Lambda_{UV}^2 + \log + \log \text{div.} + \text{finite} \quad (2.2)$$

If we compare (2.1) with (2.2) we immediately see a way to solve the hierarchy problem. If any fermion that couples to the Higgs has associated complex scalars with coupling $\lambda_S = |\lambda_f|^2$ the contributions exactly cancel due to the relative minus sign between the femionic and the bosonic loop. However, as already explained in the introduction, 't Hooft's naturalness argument forbids large accidental cancellation. Therefore, we need a symmetry which enforces this cancellation by relating bosons to fermions: supersymmetry.

Even though the cancellation of quadratic divergences is a main motivation for supersymmetry it possesses several other pleasant features which makes it maybe the best motivated, but at least most studied way to extend the SM:

- According to the Coleman Mandula theorem [45, 46] it is the only possible symmetry which can relate internal and external symmetries in a non-trivial way.

- Local supersymmetry incorporates gravity since the anticommutator of the supersymmetry-generators yields a space-time translation.

- Supersymmetry is a prediction of string theory. For consistency a world-sheet supersymmetry is required because non-supersymmetric string theories have a tachyon in their particle spectrum. Remarkably (and also maybe surprisingly) two dimensional world-sheet supersymmetry also leads to space-time supersymmetry.

- Supersymmetry improves the merging of the gauge couplings at the GUT scale. Furthermore, the unification-scale is pushed to higher energies which ensures the consistency of the predicted life-time of the proton with experiment.

2.1. Construction of the MSSM

In order to obtain a realistic supersymmetric version of the Standard Model one has to extend the field content of the theory by adding appropriate scalar or fermionic partners to the ordinary matter and gauge fields. However, one also has to extend the Higgs sector due to the analyticity of the superpotential. Analytic means that it can only be constructed as a function of fields and not of their complex conjugates. Therefore, it is impossible to generate all fermions masses using a single Higgs doublet. Instead at least two of them with opposite hypercharge are required. Therefore, the minimal field content of a realistic supersymmetrized version of the SM is:

- Vector supermultiplets containing gauge bosons and gauginos transforming in the adjoint representation of the SM gauge group $SU(3)_C \otimes SU(2)_L \otimes U(1)_Y$.

  - $V_Y \supset B_\mu, \tilde{\lambda}_B$: The U(1) hypercharge gauge field and its superpartner the bino, respectively. Both couple to matter fields with g_1
  - $V_W^k \supset W_\mu^k, \tilde{\lambda}_W^k$: The three gauge bosons associated with $SU(2)_L$ and their superpartners the winos.
  - $V_g^a \supset g_\mu^a, \tilde{g}^a$: Eight gluons and gluinos associated with $SU(3)_C$.

- Three generations (with flavor indices i, j) of chiral supermultiplets containing the SM fermions and their scalar partners the sfermions which are denoted by a tilde:

  - $SU(2)_L$ antilepton singlets: $\bar{E}_i \supset \ell_{iR}^C, \tilde{\ell}_{iR}^*$
  - $SU(2)_L$ antiquark singlets: $\bar{U}_i \supset u_{iR}^C, \tilde{u}_{iR}^*$
    $\bar{D}_i \supset d_{iR}^C, \tilde{d}_{iR}^*$
  - $SU(2)_L$ lepton doublets: $L_i = \begin{pmatrix} L_{\nu_i} \\ L_{\ell_i} \end{pmatrix} \supset \begin{pmatrix} \nu_i \\ \ell_{iL} \end{pmatrix}, \begin{pmatrix} \tilde{\nu}_i \\ \tilde{\ell}_{iL} \end{pmatrix}$

2.1 Construction of the MSSM

- $SU(2)_L$ quark doublets: $Q_i = \begin{pmatrix} U_i \\ D_i \end{pmatrix} \supset \begin{pmatrix} u_{iL} \\ d_{iL} \end{pmatrix}, \begin{pmatrix} \tilde{u}_{iL} \\ \tilde{u}_{iL} \end{pmatrix}$

- Two chiral superfields containing the two Higgs $SU(2)_L$ doublets and their fermionic superpartners, the higgsinos:

$$H_d = \begin{pmatrix} H_d^1 \\ H_d^2 \end{pmatrix} \subset \begin{pmatrix} h_d^0 \\ h_d^- \end{pmatrix}, \begin{pmatrix} \tilde{h}_d^0 \\ \tilde{h}_d^- \end{pmatrix}, \quad H_u = \begin{pmatrix} H_u^1 \\ H_u^2 \end{pmatrix} \subset \begin{pmatrix} h_u^+ \\ h_u^0 \end{pmatrix}, \begin{pmatrix} \tilde{h}_u^+ \\ \tilde{h}_u^0 \end{pmatrix}$$

The subscripts u and d anticipate to which quarks the Higgses will couple to.

In order to complete the supersymmetric part of the MSSM one has to write down the superpotential. The most general analytic gauge invariant expression which does not violate SM conservation laws is:

$$W_{MSSM} = \mu H_d \cdot H_u - Y_{ij}^\ell H_d \cdot L_i \bar{E}_j - Y_{ij}^d H_d \cdot Q_i \bar{D}_j - Y_{ij}^u H_u \cdot Q_i \bar{U}_j \tag{2.3}$$

The dot represents the $SU(2)_L$ invariant contraction of two doublets $\epsilon_{DE} A^D B^E$. The matrices in flavor space, Y_{ij}^f, are the Yukawa couplings and μ is the higgsino mass parameter. However, since no suppersymmetric particle has been discovered (yet), we know that they must be much heavier than the SM particles. Therefore, SUSY cannot be exact but rather must be broken. In order for the non-renormalization theorem to stay valid, the SUSY breaking terms must be "soft" or of positive mass dimension. Therefore, in R-parity conserving SUSY, we can write down the following soft-supersymmetry terms which respect both gauge invariance and SM conservation laws:

- Sfermion mass terms:

$$-\left(\tilde{q}_{iL}^* \left(\mathbf{M}_{LL}^{\tilde{q}\,2}\right)_{ij} \tilde{q}_{jL} + \tilde{u}_{iR}^* \left(\mathbf{M}_{RR}^{\tilde{u}\,2}\right)_{ij} \tilde{u}_{jR} + \tilde{d}_{iR}^* \left(\mathbf{M}_{RR}^{\tilde{d}\,2}\right)_{ij} \tilde{d}_{jR} \right.$$
$$\left. + \tilde{\ell}_{iL}^* \left(\mathbf{M}_{LL}^{\tilde{\ell}\,2}\right)_{ij} \tilde{\ell}_{jL} + \tilde{\ell}_{iR}^* \left(\mathbf{M}_{RR}^{\tilde{\ell}\,2}\right)_{ij} \tilde{\ell}_{jL}\right) \tag{2.4}$$

A priori, the matrices in flavor space $\mathbf{M}_{AB}^{\tilde{f}}$ are arbitrary. However, in order to obtain physical sfermions with positive mass, the diagonal entries must be bigger than the off-diagonal ones. Note that due to SU(2) invariance the mass terms for left-handed up and down squarks are equal.

- Majorana mass terms for the gaugions:

$$+\frac{1}{2}\left(M_1 \bar{\tilde{\lambda}}_0 P_L \tilde{\lambda}_0 + M_2 \bar{\tilde{\lambda}}_W^k P_L \tilde{\lambda}_W^k + m_{\tilde{g}} \bar{\tilde{g}}^a P_L \tilde{g}^a\right) + h.c. \tag{2.5}$$

- Bilinear couplings of the Higgs fields:

$$m_d^2 |h_d|^2 + m_u^2 |h_u|^2 + (B\mu h_d \cdot h_u + h.c.) \qquad (2.6)$$

- Trilinear couplings of sfermions to the Higgs fields analogous to the Yukawa terms in the superpotential:

$$h_d \cdot \tilde{\ell}_{iL} \mathbf{A}^\ell_{w\,ij} \tilde{\ell}^*_{jR} + h_d \cdot \tilde{q}_{iL} \mathbf{A}^d_{w\,ij} \tilde{d}^*_{jR} + \tilde{q}_{iL} \cdot h_u \mathbf{A}^u_{w\,ij} \tilde{u}^*_{jR} + h.c. \qquad (2.7)$$

The parameters A carry a subscript w to remind us that they are given in a weak eigenbasis. Note that these terms are not only potential sources of flavor violation, they also connect left-handed with right-handed sfermions.

- Non-analytic trilinear coupling of sfermions to the wrong Higgs field:

$$h_u^{I*} \tilde{\ell}_{iL}^I \mathbf{A}'^\ell_{w\,ij} \tilde{\ell}^*_{jR} + h_u^{I*} \tilde{q}_{iL}^I \mathbf{A}'^d_{w\,ij} \tilde{d}^*_{jR} + h_d^{I*} \tilde{q}_{iL}^I \mathbf{A}'^u_{w\,ij} \tilde{u}^*_{jR} + h.c. \qquad (2.8)$$

Even though these terms are not generated in the most popular SUSY breaking scenarios they are allowed by gauge invariance and SM conservation laws. As we will see later, in gaugino-mediated FCNCs the combination $v_d A^d$ ($v_u A^u$) cannot be distinguished from $v_u A'^d$ ($v_d A'^u$). However, the non-analytic A terms can lead to large effects in Higgs mediated FCNCs since they generate non-holomorphic Higgs couplings at the one-loop level.

2.2. The mass spectrum of the MSSM

In order to obtain the physical spectrum at low energies we have to carry out the standard procedure of spontaneous symmetry breaking. Thus, each of the Higgs fields acquires a vacuum expectation value in its neutral component:

$$\langle h_u \rangle = \begin{pmatrix} v_u \\ 0 \end{pmatrix}, \qquad \langle h_d \rangle = \begin{pmatrix} 0 \\ v_d \end{pmatrix} \qquad (2.9)$$

In this way, the $SU(2)_L$ gauge bosons receive their masses and the neutral W mixes with the U(1) gauge boson resulting in the photon and the Z. In this respect all SM formulas for the one Higgs doublet model hold with the simple replacement $v \to \sqrt{v_u^2 + v_d^2}$. The normalization in (2.9) is chosen in such a way that the (tree level) masses are given by:

$$\begin{aligned} m_{ij}^{q(0)} &= v_q Y_{ij}^q \\ m_{ij}^{\ell(0)} &= v_d Y_{ij}^\ell \end{aligned} \qquad (2.10)$$

2.2 The mass spectrum of the MSSM

We arrive at the mass eigenbasis by diagonalizing these matrices by a biunitary transformation:

$$U_L^{(0)f\dagger} m_f^{(0)} U_R^{(0)f} = m_f^{D(0)} \tag{2.11}$$

Here $m_f^{D(0)}$ is a diagonal matrix containing the singular values. Note that this relation is defined for the uncorrected tree-level masses. This will be important later, when we discuss the definition of the super-CKM basis in the presence of order one loop-corrections. With these conventions the CKM matrix is given by:

$$V^{(0)} = U_L^{(0)d\dagger} U_L^{(0)u} \tag{2.12}$$

However, we can simplify things by using our freedom to specify a weak basis. We choose:

$$U_L^{(0)d} = U_R^{(0)d} = U_R^{(0)u} = U_L^{(0)\ell} = U_R^{(0)\ell} = \hat{1}, \qquad U_L^{(0)u} = V^{(0)} \tag{2.13}$$

2.2.1. Sfermions

Using the six dimensional vector $\tilde{f} = \begin{pmatrix} \tilde{f}_L \\ \tilde{f}_R \end{pmatrix}$ as a basis, we can define the sfermion mass matrices. However, in order to consider FC processes it is very useful to switch to the super-CKM basis because in this way unphysical fermion field rotations can be absorbed into the definition of the sfermion mass matrices. We arrive at the super-CKM basis by applying the same rotations to the sfermion field which were needed to diagonalize fermion Yukawa couplings:

$$\tilde{u} = \begin{pmatrix} \tilde{u}_L \\ \tilde{u}_R \end{pmatrix} \rightarrow \begin{pmatrix} V^{(0)\dagger} \tilde{f}_L \\ \tilde{f}_R \end{pmatrix} \tag{2.14}$$

Then the sfermion mass matrices in the super-CKM basis are given by:

$$\mathbf{M}_{\tilde{u}}^2 = \tag{2.15}$$

$$\begin{pmatrix} V^{(0)\dagger} \mathbf{M}_{LL}^{\tilde{q}\,2} V^{(0)} + \frac{\cos 2\beta}{6} (m_Z^2 + 2m_W^2) \hat{1} + \left(\mathbf{m}_u^{D(0)}\right)^2 & -V^{(0)\dagger} \left(v_u \mathbf{A}_w^u + v_d \mathbf{A}_w'^u\right) + \mathbf{m}_u^{D(0)} \mu \cot\beta \\ -\left(v_u \mathbf{A}_w^{u\dagger} + v_d \mathbf{A}_w'^{u\dagger}\right) V^{(0)} - \mathbf{m}_u^{D(0)} \mu^* \cot\beta & \mathbf{M}_{RR}^{\tilde{u}\,2} + \frac{2\cos 2\beta}{3} m_Z^2 \sin^2\theta_W \hat{1} + \left(\mathbf{m}_u^{D(0)}\right)^2 \end{pmatrix}$$

$$\mathbf{M}_{\tilde{d}}^2 = \begin{pmatrix} \mathbf{M}_{LL}^{\tilde{q}\,2} - \frac{\cos 2\beta}{6} (m_Z^2 - 4m_W^2) \hat{1} + \left(\mathbf{m}_d^{D(0)}\right)^2 & -v_d \mathbf{A}_w^d - v_u \mathbf{A}_w'^d + \mathbf{m}_d^{D(0)} \mu \tan\beta \\ -v_d \mathbf{A}_w^{d\dagger} - v_u \mathbf{A}_w'^{d\dagger} - \mathbf{m}_d^{D(0)} \mu^* \tan\beta & \mathbf{M}_{RR}^{\tilde{d}\,2} - \frac{\cos 2\beta}{3} m_Z^2 \sin^2\theta_W \hat{1} + \left(\mathbf{m}_d^{D(0)}\right)^2 \end{pmatrix}$$

$$\mathbf{M}_{\tilde{\ell}}^2 = \begin{pmatrix} \mathbf{M}_{LL}^{\tilde{\ell}\,2} - \frac{\cos 2\beta}{6} (m_Z^2 - 2m_W^2) \hat{1} + \left(\mathbf{m}_\ell^{D(0)}\right)^2 & -v_d \mathbf{A}_w^\ell - v_u \mathbf{A}_w'^\ell + \mathbf{m}_\ell^{D(0)} \mu \tan\beta \\ -v_d \mathbf{A}_w^{\ell\dagger} - v_u \mathbf{A}_w'^{\ell\dagger} - \mathbf{m}_\ell^{D(0)} \mu^* \tan\beta & \mathbf{M}_{RR}^{\tilde{\ell}\,2} - \cos 2\beta m_Z^2 \sin^2\theta_W \hat{1} + \left(\mathbf{m}_\ell^{D(0)}\right)^2 \end{pmatrix}$$

Thus the trilinear terms of the super-CKM basis, A^f, and those in the weak basis are related as:
$$A^{d,\ell} = A_w^{d,\ell}, \qquad A^u = V^{(0)\dagger} A_w^u \qquad (2.16)$$
It is common to parameterize the sfermion mass matrices, given in the super-CKM basis in the following way:

$$M_{\tilde{f}}^2 = \begin{pmatrix} \left(M_{1L}^{\tilde{f}}\right)^2 & \Delta_{12}^{\tilde{f}\,LL} & \Delta_{13}^{\tilde{f}\,LL} & \Delta_{11}^{\tilde{f}\,LR} & \Delta_{12}^{\tilde{f}\,LR} & \Delta_{13}^{\tilde{f}\,LR} \\ \Delta_{12}^{\tilde{f}\,LL*} & \left(M_{2L}^{\tilde{q}}\right)^2 & \Delta_{23}^{\tilde{f}\,LL} & \Delta_{12}^{\tilde{f}\,RL*} & \Delta_{22}^{\tilde{f}\,LR} & \Delta_{23}^{\tilde{f}\,LR} \\ \Delta_{13}^{\tilde{f}\,LL*} & \Delta_{23}^{\tilde{f}\,LL*} & \left(M_{3L}^{\tilde{q}}\right)^2 & \Delta_{13}^{\tilde{f}\,RL*} & \Delta_{23}^{\tilde{f}\,RL*} & \Delta_{33}^{\tilde{f}\,LR} \\ \Delta_{11}^{\tilde{f}\,LR*} & \Delta_{12}^{\tilde{f}\,RL} & \Delta_{13}^{\tilde{f}\,RL} & \left(M_{1R}^{\tilde{q}}\right)^2 & \Delta_{12}^{\tilde{f}\,RR} & \Delta_{13}^{\tilde{f}\,RR} \\ \Delta_{12}^{\tilde{f}\,LR*} & \Delta_{22}^{\tilde{f}\,LR*} & \Delta_{23}^{\tilde{f}\,RL} & \Delta_{12}^{\tilde{f}\,RR*} & \left(M_{2R}^{\tilde{q}}\right)^2 & \Delta_{23}^{\tilde{f}\,RR} \\ \Delta_{13}^{\tilde{f}\,LR*} & \Delta_{23}^{\tilde{f}\,LR*} & \Delta_{33}^{\tilde{f}\,LR*} & \Delta_{13}^{\tilde{f}\,RR*} & \Delta_{23}^{\tilde{f}\,RR*} & \left(M_{3R}^{\tilde{q}}\right)^2 \end{pmatrix} \qquad (2.17)$$

This is useful because in gaugino mediated FCNCs always the whole off-diagonal element enters. However, the sfermion mass matrices are not (necessarily) diagonal in flavor space. Therefore, we have to diagonalize (2.17) by a unitary transformation in order to obtain the squark masses and mixing angles:
$$W^{\tilde{f}\dagger} M_{\tilde{f}}^2 W^{\tilde{f}} = M_{\tilde{f}}^{(D)2} \qquad (2.18)$$
The matrix $M_{\tilde{f}}^{(D)2}$ contains the six physical sfermion mass eigenvalues squared, $m_{\tilde{f}_s}^2$, as diagonal elements. These rotations matrices, $W_{st}^{\tilde{f}}$, which parameterize the misalignment between the squarks and the quarks enter in the squark-quark-gluino vertex (see appendix). In order to understand flavor changes induced by sfermion mass matrices qualitatively the "mass insertion approximation" is useful. This just corresponds to the expansion of (2.17) in the (assumed) small off-diagonal elements $\Delta_{ij}^{\tilde{f}\,AB}$ over the diagonal ones. Therefore, the vertices are flavor diagonal in this approximation, but flavor-changing mass terms appear on the squark lines (see Fig. 2.2). Every insertion of such a mass terms yields an additional squark propagator. This motivates the following definition for the mass insertion parameter:
$$\delta_{ij}^{f\,AB} = \frac{\Delta_{ij}^{\tilde{f}\,AB}}{M_{f\,A}^{\tilde{f}} M_{f\,B}^{\tilde{f}}} \qquad (2.19)$$
Note that even though this parameter is dimensionless, it does not only contain SUSY quantities. In the case when it is chirality flipping ($A \neq B$) it is proportional to a electroweak vacuum expectation value (vev). Therefore, theses elements do not stay constant if one scales all SUSY parameters by a common factor but decrease as v/M_{SUSY}

2.2 The mass spectrum of the MSSM

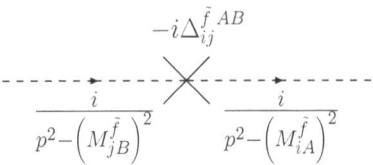

Figure 2.2: Insertion of a flavor-changing mass term into a sfermion line.

2.2.2. Charginos

The charged higgsinos and winos have the same quantum numbers and mix after electroweak symmetry breaking. The four 2-component spinors $\tilde{\lambda}_W^1, \tilde{\lambda}_W^2, \tilde{h}_d^-, \tilde{h}_u^+$ form two physical Dirac fermions $\tilde{\chi}_1^\pm, \tilde{\chi}_2^\pm$ with masses $m_1^{\tilde{\chi}^\pm}, m_2^{\tilde{\chi}^\pm}$ arising from a biunitary transformation of the chargino mass matrix:

$$U^{\tilde{\chi}^\pm *}\begin{pmatrix} M_2 & g_2 v_u \\ g_2 v_d & \mu \end{pmatrix}\left(V^{\tilde{\chi}^\pm}\right)^{-1} = \begin{pmatrix} m_1^{\tilde{\chi}^\pm} & 0 \\ 0 & m_2^{\tilde{\chi}^\pm} \end{pmatrix} \quad (2.20)$$

Interactions involving charginos are proportional to the CKM matrix (see Appendix). Therefore, charginos contribute to flavor changing processes even in the case of minimal flavor violation.

2.2.3. Neutralinos

As in the case of the charginos also the bino, the neutral wino and the neutral higgsinos share the same quantum numbers and mix among each other.

$$\left(Z^{\tilde{\chi}^0}\right)^T \begin{pmatrix} M_1 & 0 & -\frac{g_1 v_d}{\sqrt{2}} & \frac{g_1 v_u}{\sqrt{2}} \\ 0 & M_2 & \frac{g_2 v_d}{\sqrt{2}} & -\frac{g_2 v_u}{\sqrt{2}} \\ -\frac{g_1 v_d}{\sqrt{2}} & \frac{g_2 v_d}{\sqrt{2}} & 0 & -\mu \\ \frac{g_1 v_u}{\sqrt{2}} & -\frac{g_2 v_u}{\sqrt{2}} & -\mu & 0 \end{pmatrix} Z^{\tilde{\chi}^0} = \begin{pmatrix} m_1^{\tilde{\chi}^0} & 0 & 0 & 0 \\ 0 & m_2^{\tilde{\chi}^0} & 0 & 0 \\ 0 & 0 & m_3^{\tilde{\chi}^0} & 0 \\ 0 & 0 & 0 & m_4^{\tilde{\chi}^0} \end{pmatrix} \quad (2.21)$$

Note that the neutralinos possess a symmetric mass matrix. Therefore, the transposed and not the hermition conjugate of Z appears on the left side. Like the gluinos, also the neutralinos contribute only to FCNC processes (at one loop) in the case of non-minimal flavor violation.

This completes the physical content of the MSSM and we can now switch to the applications in flavor-changing processes. The simplest possible flavor-changing diagrams (with external fermions) one can construct are self-energies. We will study their effects in the following chapter.

3. Finite renormalization of fermion masses and mixing matrices

In this chapter we compute the finite renormalization of fermion masses and flavor valued wave functions induced through one-particle irreducible self-energies. We first consider the general case and then specify to the MSSM in which NLO effects become important due to a possible chiral enhancement.

3.1. General formalism

In this section we consider the general effect of finite one-particle irreducible self-energies on the mass and wave-function renormalization of fermions. It is possible to decompose any self-energy into its chirality-flipping and its chirality-conserving parts in the following way:

$$\Sigma_{ij}^f(p) = \left(\Sigma_{ij}^{f\ LR}(p^2) + \slashed{p}\Sigma_{ij}^{f\ RR}(p^2)\right) P_R + \left(\Sigma_{ij}^{f\ RL}(p^2) + \slashed{p}\Sigma_{ij}^{f\ LL}(p^2)\right) P_L . \quad (3.1)$$

Note that chirality-changing parts $\Sigma_{ij}^{f\ LR}$ and $\Sigma_{ij}^{f\ RL}$ have mass dimension 1, while $\Sigma_{ij}^{f\ LL}$ and $\Sigma_{ij}^{f\ RR}$ are dimensionless. With this convention the tree-level fermion masses

$$m_{f_i}^{(0)} = Y^{f_i\,(0)} v_{d,u} \quad (3.2)$$

receive the following corrections at the one-loop level:

$$m_{f_i}^{(0)} \to m_{f_i}^{(0)} + \Sigma_{ii}^{f\ LR}(m_{f_i}^2) + \frac{1}{2}m_{f_i}\left(\Sigma_{ii}^{f\ LL}(m_{f_i}^2) + \Sigma_{ii}^{f\ RR}(m_{f_i}^2)\right) + \delta m_{f_i} = m_{f_i}. \quad (3.3)$$

If the self-energies are finite, the counter-term δm_{f_i} in (3.3) is zero in a minimal renormalization scheme like $\overline{\text{MS}}$. Even though in a minimal renormalization scheme $m_{f_i}^{(0)}$ is not equal to the physical fermion mass we choose the $\overline{\text{MS}}$ scheme from now on for two reasons:

- Off-diagonal elements of the sfermion mass matrices, in particular the trilinear A-terms, are theoretical quantities which are not directly related to physical observables.

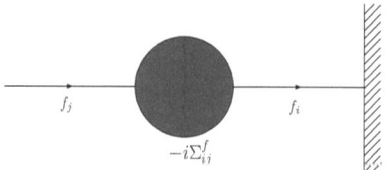

Figure 3.1: Flavour-valued wave-function renormalization.

For such quantities it is always easier to use a minimal scheme which allows for a direct relation between theoretical quantities and observables.

- We will consider the limit in which the light fermion masses and CKM elements are generated radiatively. In this limit it would be unnatural to have tree-level Yukawa couplings and CKM elements in the Lagrangian which would cancel with the counterterms in the on-shell scheme.

We will further elaborate on this important point in section 3.4 and section 5.1.
Note that m_{f_i} on both sides of equation 3.3 is the measured fermion mass in the $\overline{\text{MS}}$ scheme. Even though the fermion mass on the left-handed side arises via the equation of motion, it can be shown that taking into account the loop corrections it equals the $\overline{\text{MS}}$ mass [47]. This observation is consistent with the result in the effective field theory approach (see for example [48]).
The self-energies of equation (3.1) do not only renormalize the fermion masses. Also a rotation, $\delta_{ij} + \Delta U_{ij}^{f\,L}$, in flavor-space which has to be applied to all external fields is induced through the diagram in Fig. 3.1:

$$\Delta U_{ij}^{f\,L} = \frac{1}{m_{f_j}^2 - m_{f_i}^2} \left(m_{f_j}^2 \Sigma_{ij}^{f\,LL}\left(m_{f_i}^2\right) + m_{f_j} m_{f_i} \Sigma_{ij}^{f\,RR}\left(m_{f_i}^2\right) \right.$$
$$\left. + m_{f_j} \Sigma_{ij}^{f\,LR}\left(m_{f_i}^2\right) + m_{f_i} \Sigma_{ij}^{f\,RL}\left(m_{f_i}^2\right) \right) \quad \text{for } i \neq j, \qquad (3.4)$$

$$\Delta U_{ii}^{f\,L} = \frac{1}{2}\text{Re}\left[\Sigma_{ii}^{f\,LL}\left(m_{f_i}^2\right) + 2 m_{f_i} \Sigma_{ii}^{f\,LR\prime}\left(m_{f_i}^2\right) + m_{f_i}^2 \left(\Sigma_{ii}^{f\,LL\prime}\left(m_{f_i}^2\right) + \Sigma_{ii}^{f\,RR\prime}\left(m_{f_i}^2\right) \right) \right].$$

The prime denotes differentiation with respect to the argument. The flavor-diagonal part arises from the truncation of flavor-conserving self-energies. The equations (3.3) and (3.4) are valid for arbitrary one-particle irreducible self-energies and we are going to apply them to the MSSM after we have calculated the supersymmetric self-energies in the next section.

3.2. Self-energies in the MSSM

Self-energies with supersymmetric virtual particles are of special importance because of a possible chiral enhancement which can lead to order-one corrections. In this section we calculate the chirally enhanced (by a factor $\frac{A_f^{ij}}{M_{SUSY} Y_f^{ij}}$ or $\tan\beta$) parts of the fermion self-energies in the MSSM. For this purpose it is sufficient to evaluate the diagrams at vanishing external momentum.

We choose the sign of the self-energies Σ to be equal to the sign of the mass, e.g. calculating a self-energy diagram yields $-i\Sigma$. Then, with the Feynman rules given in the appendix, the fermion self-energies with a gaugino and a sfermion as virtual particles are given by:

$$\begin{aligned}
\Sigma_{f_i-f_j}^{\tilde{\lambda}\,LR} &= \frac{-1}{16\pi^2} \sum_{s=1}^{6} \sum_{k=1}^{l} m_{\tilde{\lambda}_k} \Gamma_{f_i\tilde{f}_s}^{\tilde{\lambda}_k L*} \Gamma_{f_i\tilde{f}_s}^{\tilde{\lambda}_k R} B_0\left(p^2; m_{\tilde{\lambda}_k}^2, m_{\tilde{f}_s}^2\right) \\
\Sigma_{f_i-f_j}^{\tilde{\lambda}\,RL} &= \frac{-1}{16\pi^2} \sum_{s=1}^{6} \sum_{k=1}^{l} m_{\tilde{\lambda}_k} \Gamma_{f_i\tilde{f}_s}^{\tilde{\lambda}_k R*} \Gamma_{f_i\tilde{f}_s}^{\tilde{\lambda}_k L} B_0\left(p^2; m_{\tilde{\lambda}_k}^2, m_{\tilde{f}_s}^2\right) \\
\Sigma_{f_i-f_j}^{\tilde{\lambda}\,LL} &= \frac{-1}{16\pi^2} \sum_{s=1}^{6} \sum_{k=1}^{l} \Gamma_{f_i\tilde{f}_s}^{\tilde{\lambda}_k L*} \Gamma_{f_i\tilde{f}_s}^{\tilde{\lambda}_k L} B_1\left(p^2; m_{\tilde{\lambda}_k}^2, m_{\tilde{f}_s}^2\right) \\
\Sigma_{f_i-f_j}^{\tilde{\lambda}\,RR} &= \frac{-1}{16\pi^2} \sum_{s=1}^{6} \sum_{k=1}^{l} \Gamma_{f_i\tilde{f}_s}^{\tilde{\lambda}_k R*} \Gamma_{f_i\tilde{f}_s}^{\tilde{\lambda}_k R} B_1\left(p^2; m_{\tilde{\lambda}_k}^2, m_{\tilde{f}_s}^2\right)
\end{aligned} \quad (3.5)$$

Here $\tilde{\lambda}$ stands for the gauginos ($\tilde{g}, \tilde{\chi}^0, \tilde{\chi}^\pm$) and l denotes their corresponding number (2 for charginos, 4 for neutralinos and 8 for gluinos). The loop functions B_0 and B_1 are defined as:

$$\begin{aligned}
B_0\left(p^2; m_1^2, m_2^2\right) &= \frac{(2\pi\mu)^{4-d}}{i\pi^2} \int d^d k \frac{1}{k^2 - m_1^2} \frac{1}{(k-p)^2 - m_2^2} \\
p^\mu B_1\left(p^2; m_1^2, m_2^2\right) &= \frac{(2\pi\mu)^{4-d}}{i\pi^2} \int d^d k \frac{k^\mu}{k^2 - m_1^2} \frac{1}{(k-p)^2 - m_2^2}
\end{aligned} \quad (3.6)$$

Note that the self-energies in equation (3.5) are both finite and independent of the renormalization scale if the flavor or chirality change stems from the sfermion mass matrix. Since we know from experiment that the SUSY particles must be much heavier than the five lightest quarks we can expand the functions in equation (3.6) in the external momentum p. Using

$$\frac{1}{(k-p)^2 - m^2} = \frac{1}{k^2 - m^2} + \frac{2k^\mu p_\mu}{(k^2 - m^2)^2} + \frac{p^2 m^2 + k^\mu k^\nu (4p_\mu p_\nu - p^2 g_{\mu\nu})}{(k^2 - m^2)^3} \\
+ \frac{-4\left((k^\mu k^\nu k^\rho (g_{\mu\nu} p^2 - 2p_\mu p_\nu) - k^\rho m^2 p^2)p_\rho\right)}{(k^2 - m^2)^4} + ... \quad (3.7)$$

we obtain:

$$B_0\left(p^2; m_1^2, m_2^2\right) = B_0\left(m_1^2, m_2^2\right) + p^2 m_2^2 D_0\left(m_1^2, m_2^2, m_2^2, m_2^2\right) + \ldots$$
$$B_1\left(p^2; m_1^2, m_2^2\right) = \frac{1}{2}C_2\left(m_1^2, m_2^2, m_2^2\right) + m^2 p^2 E_2\left(m_1^2, m_2^2, m_2^2, m_2^2, m_2^2\right) + \ldots \quad (3.8)$$

The explicit formula for the loop-functions are given in the appendix. As already noted in the introduction the chirally enhanced parts, which just correspond to vanishing external momentum, are of special importance. We give explicit results for these part below since we will need them for our numerical analysis in chapter 4 and 5. First of all, the quark self-energy with a gluino and a squark as virtual particles is given by

$$\Sigma_{q_i q_j}^{\tilde{g}\,LR} = \frac{2\alpha_s}{3\pi} m_{\tilde{g}} \sum_{s=1}^{6} W_{(j+3)s}^{\tilde{q}*} W_{is}^{\tilde{q}} B_0(m_{\tilde{g}}^2, m_{\tilde{q}_s}^2) \quad (3.9)$$

For the fermion self-energy with a neutralino we get:

$$\begin{aligned}\Sigma_{f_i f_j}^{\tilde{\chi}^0\,LR} = \frac{-1}{16\pi^2}\sum_{s=1}^{6}\sum_{k=1}^{m} m_{\tilde{\chi}_k^0} &\left(-\left(a_1^f g_1 g_2 Z_{k2}^{\tilde{\chi}^0} + a_2^f g_1^2 Z_{k1}^{\tilde{\chi}^0}\right) Z_{k1}^{\tilde{\chi}^0} W_{is}^{\tilde{f}} W_{j+3,s}^{\tilde{f}*} \right.\\ &- \left(a_3^f \frac{g_1}{\sqrt{2}} Z_{k1}^{\tilde{\chi}^0} + a_4^f \frac{g_2}{\sqrt{2}} Z_{k2}^{\tilde{\chi}^0}\right) Y^{f_i} Z_{k3}^{\tilde{\chi}^0} W_{is}^{\tilde{f}} W_{js}^{\tilde{f}*} \\ &+ a_5^f \sqrt{2} g_1 Y^{f_i} Z_{k3}^{\tilde{\chi}^0} Z_{k1}^{\tilde{\chi}^0} W_{i+3,s}^{\tilde{f}} W_{j+3,s}^{\tilde{f}*} \\ &\left.+ \left(Y^{f_i} Z_{k3}^{\tilde{\chi}^0}\right)^2 W_{i+3,s}^{\tilde{f}} W_{js}^{\tilde{f}*}\right) B_0\left(m_{\tilde{\chi}^0}^2, m_{\tilde{d}_s}^2\right)\end{aligned} \quad (3.10)$$

The coefficients a^f are one for the leptons and in the case of down-type and up-type quarks they are given by:

$$\begin{array}{lllll} a_1^d = \frac{1}{3} & a_2^d = -\frac{2}{9} & a_3^d = -\frac{1}{3} & a_4^d = 1 & a_5^d = \frac{1}{3} \\ a_1^u = \frac{2}{3} & a_2^u = \frac{2}{9} & a_3^u = \frac{1}{3} & a_4^u = -1 & a_5^u = \frac{-2}{3} \end{array} \quad (3.11)$$

Finally we receive for the down-quark and lepton self-energy with a chargino:

$$\Sigma_{d_i d_j}^{\tilde{\chi}^\pm\,LR} = \frac{-1}{16\pi^2}\sum_{s=1}^{6}\sum_{k=1}^{2}\sum_{m,n=1}^{3}\left(m_{\tilde{\chi}_k^\pm}\left(V_{k2}^{\tilde{\chi}^\pm}Y^{u_m*}Y^{d_j}W_{m+3,s}^{\tilde{u}} - g_2 V_{k1}^{\tilde{\chi}^\pm}Y^{d_j}W_{ms}^{\tilde{u}}\right)\right.$$
$$\left.U_{k2}^{\tilde{\chi}^\pm}V_{mi}^{CKM*}V_{nj}^{CKM}W_{ns}^{\tilde{u}*}B_0\left(m_{\tilde{\chi}_k^\pm}^2, m_{\tilde{u}_s}^2\right)\right) \quad (3.12)$$

$$\Sigma_{\ell_i \ell_j}^{\tilde{\chi}^\pm\,LR} = \frac{1}{16\pi^2}\sum_{s=1}^{6}\sum_{k=1}^{2}\sum_{m,n=1}^{3} m_{\tilde{\chi}_k^\pm} g_2 V_{k1}^{\tilde{\chi}^\pm} U_{k2}^{\tilde{\chi}^\pm} Y^{\ell_j} V_{mi}^{CKM*} V_{nj}^{CKM} W_{ms}^{\tilde{u}} W_{ns}^{\tilde{u}*} B_0\left(m_{\tilde{\chi}_k^\pm}^2, m_{\tilde{\nu}_s}^2\right) \quad (3.13)$$

The corresponding expression for up-type quarks is easily obtained by interchanging u and d. We denote the sum of all contribution as:

$$\Sigma_{ij}^{q\,LR} = \Sigma_{q_iq_j}^{\tilde{g}\,LR} + \Sigma_{q_iq_j}^{\tilde{\chi}^0\,LR} + \Sigma_{q_iq_j}^{\tilde{\chi}^\pm\,LR} \qquad (3.14)$$

$$\Sigma_{ij}^{\ell\,LR} = \Sigma_{\ell_i\ell_j}^{\tilde{\chi}^0\,LR} + \Sigma_{\ell_i\ell_j}^{\tilde{\chi}^\pm\,LR}. \qquad (3.15)$$

These self-energies can lead to significant quantum corrections to fermion masses, but except for the gluino, the pure bino ($\propto g_1^2$) and the negligible small bino-wino mixing ($\propto g_1 g_2$) contribution, they are proportional to at least one power of a tree-level Yukawa coupling. However, if the light fermion masses are generated radiatively from chiral flavor-violation in the soft SUSY-breaking terms, then the Yukawa couplings of the first and second generation even vanish and the latter effect is absent at all. Radiatively generated fermion mass terms via soft tri-linear A-terms correspond to the upper bound found from the fine-tuning argument where the correction to the mass is as large as the measured physical mass itself. This fine-tuning argument is based on 't Hooft's naturalness principle (see chapter 4 for details) and we will use it to constrain the soft SUSY breaking parameters. If we restrict ourself to the case with vanishing first and second generation tree-level Yukawa couplings, the off-diagonal entries in the sfermion mass matrices stem from the soft tri-linear terms. Thus we are left with $\delta_{ij}^{f\,LR}$ only. In the mass insertion approximation with only one LR insertion the expressions for flavor violating self-energies simplify. For the gluino (neutralino) self-energies which are relevant for our following discussion for the quark (lepton) case we get:

$$\Sigma_{q_iq_j}^{\tilde{g}\,AB} = \frac{2\alpha_s}{3\pi} M_{\tilde{g}} m_{\tilde{q}_{jB}} m_{\tilde{q}_{iA}} \delta_{ij}^{q\,AB} C_0\left(m_{\tilde{g}}^2, m_{\tilde{q}_{jB}}^2, m_{\tilde{q}_{iA}}^2\right), \qquad (3.16)$$

$$\Sigma_{\ell_i\ell_j}^{\tilde{B}\,AB} = \frac{\alpha_1}{4\pi} M_1 m_{\tilde{l}_{jB}} m_{\tilde{l}_{iA}} \delta_{ij}^{\ell\,AB} C_0\left(M_1^2, m_{\tilde{\ell}_{jB}}^2, m_{\tilde{\ell}_{iA}}^2\right). \qquad (3.17)$$

Since the sneutrino mass matrix consists only of a LL block, there are no chargino diagrams in the lepton case with LR insertions at all.

3.3. Mass and wave function renormalization in the MSSM

Since the SUSY particles are known to be much heavier than the five lightest quarks it is possible to evaluate the one-loop self-energies at vanishing external momentum and to neglect higher terms which are suppressed by powers of $m_{f_i}^2/M_{SUSY}^2$. (We will not need the chirality conserving part of the self-energies until we discuss the renormalization of the W

vertex in chapter 8.) Therefore (3.1) simplifies to

$$\Sigma_{ij}^{f\,(1)} = \Sigma_{ij}^{f\,LR\,(1)} P_R + \Sigma_{ij}^{f\,RL\,(1)} P_L \tag{3.18}$$

at the one-loop level (indicated by the superscript (1)). In this approximation the self-energies are always chirality changing and contribute to the finite renormalization of the quark masses in (3.3) and to the flavor-valued wave-function renormalization in (3.4). At the one-loop level we receive the well known result

$$m_{f_i}^{(0)} \to m_{f_i}^{(1)} = m_{f_i}^{(0)} + \Sigma_{ii}^{f\,LR\,(1)} \tag{3.19}$$

for the mass renormalization in the $\overline{\text{MS}}$ scheme. According to (3.4) the flavor-valued rotation which has to be applied to all external fermion fields is given by:

$$\Delta U^{f\,L\,(1)} = \tag{3.20}$$

$$\begin{pmatrix} 0 & \dfrac{m_{f_2}\Sigma_{12}^{f\,LR\,(1)} + m_{f_1}\Sigma_{12}^{f\,RL\,(1)}}{m_{f_2}^2 - m_{f_1}^2} & \dfrac{m_{f_3}\Sigma_{13}^{f\,LR\,(1)} + m_{f_1}\Sigma_{13}^{f\,RL\,(1)}}{m_{f_3}^2 - m_{f_1}^2} \\ \dfrac{m_{f_1}\Sigma_{21}^{f\,LR\,(1)} + m_{f_2}\Sigma_{21}^{f\,RL\,(1)}}{m_{f_1}^2 - m_{f_2}^2} & 0 & \dfrac{m_{f_3}\Sigma_{23}^{f\,LR\,(1)} + m_{f_2}\Sigma_{23}^{f\,RL\,(1)}}{m_{f_3}^2 - m_{f_2}^2} \\ \dfrac{m_{f_1}\Sigma_{31}^{f\,LR\,(1)} + m_{f_3}\Sigma_{31}^{f\,RL\,(1)}}{m_{f_1}^2 - m_{f_3}^2} & \dfrac{m_{f_2}\Sigma_{32}^{f\,LR\,(1)} + m_{f_3}\Sigma_{32}^{f\,RL\,(1)}}{m_{f_2}^2 - m_{f_3}^2} & 0 \end{pmatrix}.$$

The corresponding corrections to the right-handed wave-functions are obtained by simply exchanging L with R and vice versa in (3.20). Note that the contributions of the self-energies $\Sigma_{ij}^{f\,LR\,(1)}$ with $i > j$ are suppressed by small mass-ratios. Therefore, the corresponding off-diagonal elements of the sfermion mass matrices $\Delta_{ij}^{f\,LR\,(1)}$ cannot be constrained from the CKM and PMNS renormalization.

However, since we treat, in the spirit of Ref. [29], all diagrams in which no flavor appears twice on fermion lines as one-particle irreducible, chirally-enhanced self-energies can also be constructed at the two-loop level (see Fig. (3.2)):

$$\Sigma_{ij}^{f\,RR\,(2)}(p^2) = \sum_{k\neq i,j} \frac{\Sigma_{ik}^{f\,RL\,(1)} \Sigma_{kj}^{f\,LR\,(1)}}{p^2 - m_{f_k}^2}, \qquad \Sigma_{ij}^{f\,LL\,(2)}(p^2) = \sum_{k\neq i,j} \frac{\Sigma_{ik}^{f\,LR\,(1)} \Sigma_{kj}^{f\,RL\,(1)}}{p^2 - m_{f_k}^2},$$

$$\Sigma_{ij}^{f\,LR\,(2)}(p^2) = \sum_{k\neq i,j} m_{f_k} \frac{\Sigma_{ik}^{f\,LR\,(1)} \Sigma_{kj}^{f\,LR\,(1)}}{p^2 - m_{f_k}^2}, \qquad \Sigma_{ij}^{f\,RL\,(2)}(p^2) = \sum_{k\neq i,j} m_{f_k} \frac{\Sigma_{ik}^{f\,RL\,(1)} \Sigma_{kj}^{f\,RL\,(1)}}{p^2 - m_{f_k}^2}.$$

$$\tag{3.21}$$

3.3 Mass and wave function renormalization in the MSSM

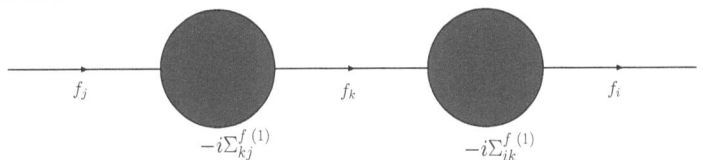

Figure 3.2: One-particle irreducible two-loop self-energy constructed out of two one-loop self-energies with $i \neq j \neq k$.

Therefore, the chirally-enhanced two-loop corrections to the masses and the wave-function renormalization are given by:

$$\begin{pmatrix} m_{f_1}^{(0)} \\ m_{f_2}^{(0)} \\ m_{f_3}^{(0)} \end{pmatrix} \to \begin{pmatrix} m_{f_1}^{(0)} + \Sigma_{11}^{f\,LR\,(1)} - \dfrac{\Sigma_{12}^{f\,LR\,(1)} \Sigma_{21}^{f\,LR\,(1)}}{m_{f_2}} - \dfrac{\Sigma_{13}^{f\,LR\,(1)} \Sigma_{31}^{f\,LR\,(1)}}{m_{f_3}} \\ m_{f_2}^{(0)} + \Sigma_{22}^{f\,LR\,(1)} - \dfrac{\Sigma_{23}^{f\,LR\,(1)} \Sigma_{32}^{f\,LR\,(1)}}{m_{f_3}} \\ m_{f_3}^{(0)} + \Sigma_{33}^{f\,LR\,(1)} \end{pmatrix}, \quad (3.22)$$

$$\Delta U_L^{f\,(2)} = \quad (3.23)$$

$$\begin{pmatrix} -\dfrac{\left|\Sigma_{12}^{f\,LR\,(1)}\right|^2}{2m_{f_2}^2} - \dfrac{\left|\Sigma_{13}^{f\,LR\,(1)}\right|^2}{2m_{f_3}^2} & -\dfrac{\Sigma_{13}^{f\,LR\,(1)} \Sigma_{32}^{f\,LR\,(1)}}{m_{f_2} m_{f_3}} & \dfrac{\Sigma_{12}^{f\,LR\,(1)} \Sigma_{23}^{f\,RL\,(1)}}{m_{f_3}^2} \\ \dfrac{\Sigma_{23}^{f\,RL\,(1)} \Sigma_{31}^{f\,RL\,(1)}}{m_{f_2} m_{f_3}} & -\dfrac{\left|\Sigma_{23}^{f\,LR\,(1)}\right|^2}{2m_{f_3}^2} - \dfrac{\left|\Sigma_{12}^{f\,LR\,(1)}\right|^2}{2m_{f_2}^2} & \dfrac{\Sigma_{21}^{f\,LR\,(1)} \Sigma_{13}^{f\,RL\,(1)}}{m_{f_3}^2} \\ \dfrac{\Sigma_{32}^{f\,RL\,(1)} \Sigma_{21}^{f\,RL\,(1)}}{m_{f_2} m_{f_3}} & -\dfrac{\Sigma_{31}^{f\,RL\,(1)} \Sigma_{12}^{f\,LR\,(1)}}{m_{f_2} m_{f_3}} & -\dfrac{\left|\Sigma_{13}^{f\,LR\,(1)}\right|^2}{2m_{f_3}^2} - \dfrac{\left|\Sigma_{23}^{f\,LR\,(1)}\right|^2}{2m_{f_3}^2} \end{pmatrix},$$

where we have neglected small mass ratios. In the quark case, we already know about the hierarchy of the self-energies from our fine-tuning argument. In this case (3.23) is just necessary to account for the unitarity of the CKM matrix as we will see in the next section. However, the corrections to $m_{f_1}^{(0)}$ in (3.22) can be large. For this reason we can also constrain $\Sigma_{31}^{f\,LR\,(1)}$ with 't Hooft's naturalness criterion if at the same time $\Sigma_{13}^{f\,LR\,(1)}$ is different from zero.

In the decoupling limit our diagrammatic method is equivalent to the effective field theory approach illustrated in Fig. 3.3. Diagonalizing (pertubatively) the effective Yukawa cou-

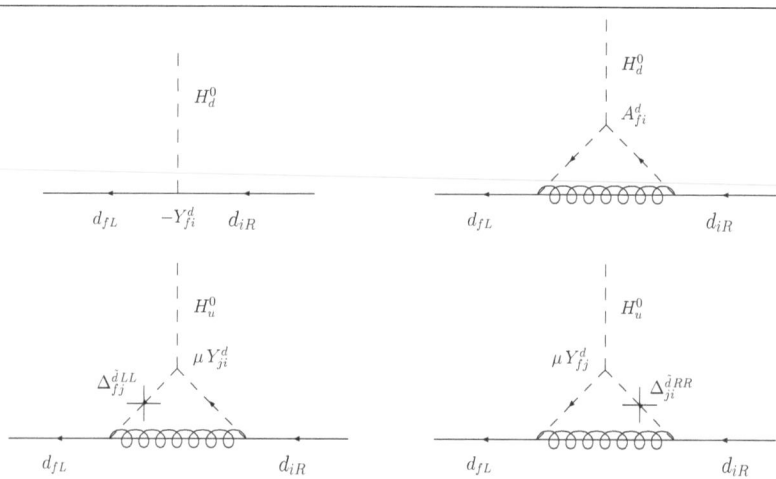

Figure 3.3: Tree–level coupling with Y_{ij}^d and FCNC loop corrections with A_{fi}^d (upper row) and $\Delta_{fi}^{\tilde{d}LL,RR}$ (lower row) in the mass insertion approximation for $M_{\text{SUSY}} \gg v$. Replacing the Higgs fields by their vevs gives the contributions to the down–type quark mass matrix. The lower diagrams contribute to the mass matrix with an enhancement factor of $\tan\beta = v_u/v_d$ compared to the other two contributions.

pling with a biunitary transformation, the rotations necessary for this diagonalization are given by $1 + \Delta U_{L,R}^{f\,(1)} + \Delta U_{L,R}^{f\,(2)}$.

Since inverse quark masses enter equation (3.20) and (3.23), we must address the proper definition of these masses in the presence of SQCD and ordinary QCD corrections. If we worked in the decoupling limit and calculated the diagrams of Fig. 3.3, we would encounter the $\overline{\text{MS}}$-renormalized Yukawa couplings evaluated at the renormalization scale $Q = M_{\text{SUSY}}$, at which the heavy SUSY particles are integrated out. Translating that result into the language of our diagrammatic approach amounts to the evaluation of the inverse quark masses in the $\overline{\text{MS}}$ scheme at $Q = M_{\text{SUSY}}$. One can derive this (somewhat surprising) result by studying QCD corrections to the diagrams of Fig. 3.5 [47]. The first element in this proof is the observation that e.g. $\Sigma_{fi}^{q\,LR}$, viewed as the Wilson coefficient of the two-quark operator $\bar{q}_f P_R q_i$, renormalizes in the same way as the quark mass, so that the

3.4 Renormalization of the CKM matrix

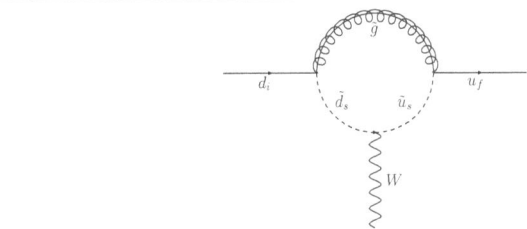

Figure 3.4: Genuine SQCD vertex correction

ratios $\Sigma_{fi}^{q\,LR}/m_{q_i}$ in (3.20) are independent of Q. Since the SUSY parameters entering Σ_{fi} are defined at the high scale $Q = M_{\text{SUSY}}$, our constraints derived in the next chapter will involve $m_{q_i}(M_{\text{SUSY}})$. The second element in the proof is the explicit analysis of gluonic corrections to the diagrams of Fig. 3.5. While at intermediate steps a quark pole mass enters through the Dirac equation $\not{p} q_i = m_{q_i}^{\text{pole}} q_i$, gluonic self-energies add to $m_{q_i}^{\text{pole}}$ in such a way that the final result only involves the properly defined $\overline{\text{MS}}$ mass m_{q_i} [47]. Note that this observation is independent of the renormalization scheme used for the supersymmetric corrections to the quark masses. Also we we treat the latter ones in the $\overline{\text{MS}}$ scheme the propagator always contains the sum $v_q Y^{q_i} + \Sigma_{ii}^{q,LR} = m_{q_i}^{\overline{\text{MS}}}$ due to the Dyson resummation.

3.4. Renormalization of the CKM matrix

In this section we calculate the renormalization of the CKM matrix in the MSSM with generic sources of flavour violation. There are two possible contributions, the self-energy diagrams of Fig. 3.5, which of course correspond to the flavour-valued wave-function renormalization of (3.20), and the proper vertex correction shown in Fig. 3.4. In the limit $M_{\text{SUSY}} \gg v$ the self-energy contributions reproduce the results of the diagrams in Fig. 3.3. (For a discussion of this feature in the MFV case see Refs. [49, 50].)

From previous considerations we know that we need some parametric enhancement (by e.g. a factor of $|A_{fi}^q|/(M_{\text{SUSY}}|Y_{fi}^q|) \gg 1$) to compensate the loop suppression and the diagrams of Fig. 3.5 involve such enhancement factors. The vertex diagram involving a W coupling to squarks is not enhanced and moreover suffers from gauge cancellations with non-enhanced pieces from the self-energies. However, as we will discuss in Chapter 7, a sizable right-handed W coupling can be induced.

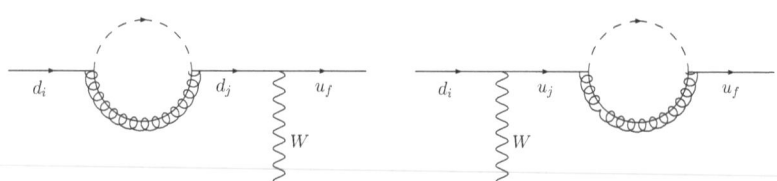

Figure 3.5: One-loop corrections to the CKM matrix from the down and up sectors.

Therefore, for the renormalization of the CKM matrix we only need to consider self-energies, just as in the case of the electroweak renormalization of the CKM matrix in the SM [51–55].

3.4.1. Super-CKM basis beyond tree-level

Since we work beyond tree-level, we have to clarify how we define the super-CKM basis in the presence of radiative corrections. Recall from chapter 2 that starting from some weak basis with Yukawa matrices Y^d and Y^u we perform the usual rotations in flavour space in equation (2.11) to diagonalize Y^q and the tree-level mass matrices $m_q^{(0)} = Y^q v_q$ and apply the same rotations to $\tilde{d}_{L,R}$ and $\tilde{u}_{L,R}$. This defines the super-CKM basis in which the elements of $M_{\tilde{q}}^2$ in (2.17) are defined. We also recall that the tree-level CKM matrix is then given by equation (2.12):

$$V^{(0)} = U_L^{(0)\,u\dagger} U_L^{(0)\,d} \qquad (3.24)$$

To fix the relation between $V^{(0)}$ and the physical CKM matrix V we must define a renormalization scheme. First note that all radiative corrections discussed in this chapter are finite, so that the notion of minimal renormalization means that all counter-terms are simply equal to zero. Two possibilities come to mind:

i) Minimal renormalization of V: The Lagrangian contains diagonal Yukawa matrices and $V^{(0)}$ without counter-terms, while the measured CKM matrix V differs from $V^{(0)}$ by the radiative corrections in Fig. 3.5. Recall that for $m_j \neq m_i$ one can treat the diagrams of Fig. 3.5 in the same way as genuine vertex corrections, i.e. there is no need to truncate such diagrams or to introduce matrix-valued wave function counter-terms [29].

3.4 Renormalization of the CKM matrix

ii) On-shell renormalization of V: The Lagrangian contains finite counter-terms to cancel the flavour-changing self-energies of Fig. 3.5. These counter-terms arise from a perturbative unitary rotation of the quark fields in flavour space, $q_{L,R} \to [1 + \delta U^q_{L,R}]q_{L,R}$ [52]. This in turn induces a counter-term

$$\delta V = \delta U_L^{u\dagger} V^{(0)} + V^{(0)} \delta U_L^d \qquad (3.25)$$

to the CKM matrix. In the on-shell scheme we can identify $V = V^{(0)}$, but after the extra rotation of the quark fields we are no more in the super-CKM basis and the bare Yukawa matrices Y^d and Y^u are no more diagonal.[1]

We choose method i), because it involves the super-CKM basis, so that we can immediately use the $\Delta^{\tilde{q}XY}_{ij}$'s defined in (2.17), permitting a direct comparison with FCNC analyses. This issue of the definition of $\Delta^{\tilde{q}XY}_{ij}$ formally goes beyond the one-loop order, but is numerically highly relevant, because the tree-level elements $V^{(0)}_{ij}$ and the finite counter-terms $[\delta U^q_L]_{ij}$ are similar in size: If one works in an alternative basis in which the (s)quark superfields are rotated by $[1 + \delta U^q_{L,R}]$, the off-diagonal elements $\Delta^{\tilde{q}XY}_{ij}$ of the squark mass matrices can substantially differ from those of our definition of the super-CKM basis.

In our super-CKM scheme i) the inclusion of the radiative correction is equivalent to the use of the tree-level coupling in the $\bar{u}W^+d$ vertex with the replacement

$$V^{(0)} \to V = \left(1 + \Delta U_L^{u\dagger}\right) V^{(0)} \left(1 + \Delta U_L^d\right) \qquad (3.26)$$

and we identify V with the physical CKM matrix. In the on-shell scheme ii) the counter-terms $\delta U_L^d = -\Delta U_L^d$ and $\delta U_L^u = -\Delta U_L^u$ cancel the loops and $V^{(0)} = V$ is maintained. It is crucial that $1 + \Delta U_L^q$ is unitary, otherwise the unitarity of V (and electroweak gauge invariance) would be spoiled [51–55]. To our one-loop order this means that ΔU_L^q is anti-hermitian. We can easily verify from (3.20) that ΔU_L^d fulfills this criterion owing to $\Sigma^{q\,RL}_{fi}(0) = [\Sigma^{q\,LR}_{if}(0)]^*$.

The self-energies do not decouple for $M_{\text{SUSY}} \to \infty$ and, in accordance with the decoupling theorem [58], we find that their mere effect is the renormalisation of the CKM matrix, as implemented in scheme ii).

It is important to stress that the replacement rule in (3.26) only absorbs the effects of the self-energy diagrams of Fig. 3.5 correctly, if both quark lines are external lines. If some

---
[1]That is, in our Feynman diagrammatic approach the FCNC Higgs couplings of Refs. [17, 48, 56, 57] enter the Lagrangian through a finite FCNC counter-term to Yukawa couplings.

$\bar{u}_f W^+ d_i$ vertex appears in a loop diagram, one or both self-energies are probed off-shell and one must work with $V^{(0)}$ and must calculate the loop diagram with the nested self-energy explicitly.

We can now understand how to treat self-energies with a top quark in (3.20): If the top quark appears on the internal line of the right diagram in Fig. 3.5, that is $j = 3$, the self-energy involved must be evaluated at $p^2 = 0$, because the external quark is up or charm. The unitarity of V now forces us to evaluate $\Sigma_{31}^{u\,RL}$ and $\Sigma_{32}^{u\,RL}$ at $p^2 = 0$ as well. Interestingly, from today's precision data in K and B physics one can determine V from tree-level data only [59]. Of course, none of these measurements involves top decays, so that the values of V_{ts} and V_{td} inferred from these measurements (through unitarity of V) indeed correspond to the definition in (3.26), with self-energies $\Sigma_{3i}^{u\,RL}$ evaluated at $p^2 = 0$. While FCNC processes of K and B mesons involve V_{ts} or V_{td}, we cannot determine these CKM elements from FCNC processes in a model-independent way, because new particles (in our case squarks and gluinos) will affect the FCNC loops directly. Clearly nothing can be learned from measuring the $\bar{t}W^+ d_i$ couplings (in, for instance, single top production or top decays) if $M_{\text{SUSY}} \gg m_{u_3} = m_t$. However, if $m_t \sim M_{\text{SUSY}}$ any on-shell $t \to s$ or $t \to d$ transition involves

$$\Delta \sigma_i^t \equiv \frac{\Sigma_{3i}^{u\,RL}(m_t^2) - \Sigma_{3i}^{u\,RL}(0)}{m_t} \qquad \text{with } i = 1 \text{ or } 2. \qquad (3.27)$$

Here the first self-energy enters the calculated $\bar{t}W^+ d_i$ process explicitly, while $\Sigma_{3i}^{u\,RL}(0)$ stems from the relationship between V and $V^{(0)}$. $\Delta \sigma_i^t$ decouples as m_t^2/M_{SUSY}^2, but can be sizable for $\mathcal{O}(200 \text{ GeV})$ superpartners, since it involves poorly-constrained FCNC squark mass terms. We conclude that the flavour structure of tree-level top couplings can help to study new physics entering chirality-flipping self-energies, while this effect is unobservable in charged-current processes of light quarks: Here the chirality-flipping self-energies merely renormalise the CKM matrix; the physical effect in a charged-current process with external quark q is suppressed by a factor of m_q^2/M_{SUSY}^2. The experimental signature would be an apparent violation of CKM unitarity, since the measured value of V_{ts} or V_{td} would be in disagreement with the value inferred from CKM unitarity. Unitarity is restored, once the correction $\Delta \sigma_i^t$ is taken into account.

We close this section by recalling the relationship between the Yukawa matrices $\mathbf{Y}^q = \text{diag}(Y^{q_1}, Y^{q_2}, Y^{q_3})$ and the quark masses [49, 50]:

$$Y^{q_i} = \frac{m_{q_i}}{v_q(1 + \Delta_{q_i})} = \frac{m_{q_i} - \Sigma_{ii,A}^{q\,LR}}{v_q\left(1 + \frac{\Sigma_{ii,\mu}^{q\,LR}}{m_{q_i}}\right)}. \qquad (3.28)$$

3.4 Renormalization of the CKM matrix

In (3.28) we have used the fact that Σ_{ii}^{qLR} can be decomposed into $\Sigma_{ii,A}^{qLR} + \Sigma_{ii,\mu}^{qLR}$ if the physical squark masses are chosen as input parameters. $\Sigma_{ii,\mu}^{qLR}$ is proportional to μY^{q_i} and $\Sigma_{ii,A}^{qLR}$ is proportional to A_{ii}^q. If we neglect the A-terms (3.28) reduces to the expression of [50] for down-type quarks. For a detailed discussion of the relation between the Yukawa matrices and the quark masses with different choices of input parameters see [47]. (3.28) holds in the super-CKM scheme i), which has the advantage that no FCNC Yukawa couplings occur. In the on-shell scheme ii) the rotations of the quark fields in flavour space lead to the loop-induced finite FCNC Yukawa couplings of Refs. [17, 48 56, 57]. In the super-CKM scheme these effects are reproduced from diagrams with flavour-diagonal Yukawa couplings and FCNC self-energies. Finally note that Δ_{q_i} can be complex, so that the entries of $\mathbf{Y}^q$ (and $m_q^{(D)} = \mathbf{Y}^q v_q$ entering the squark mass matrices in (2.15)) are not necessarily real.

3.4.2. The CKM matrix in charged-Higgs and chargino couplings

CKM elements do not only enter the Feynman rules for W couplings but also appear in the couplings of charged Higgs bosons and charginos. The Feynman rules in the super-CKM scheme i) involve $V^{(0)} = U_L^{u(0)\dagger} U_L^{d(0)}$ throughout as described in section 2. Whenever a charged Higgs boson or a chargino couples to an external quark there are chirally enhanced one-loop corrections similar to those in Figs. 3.5. We can include these diagrams by working with the tree-level diagrams and replacing $U_L^{q(0)}$ by

$$U_L^q = U_L^{q(0)}(1 + \Delta U_L^q), \qquad (3.29)$$

if the external quark is left-handed. For instance, we have shown that the loop corrections to the $\bar{u}_f W^+ d_i$ coupling were correctly included by this replacement (see (3.26)). That is, in the case of $\bar{u}_f W^+ d_i$ coupling one simply uses the physical CKM matrix V_{fi} instead of the tree-level CKM matrix $V_{fi}^{(0)}$. One immediately notices that (in the super-CKM scheme) the $\tilde{u}_f^* W^+ \tilde{d}_i$ coupling still involves $V_{fi}^{(0)}$, because the supersymmetric analogues of the diagrams of Fig. 3.5 are not chirally enhanced and will only lead to small corrections of the typical size of ordinary loop corrections. Enhanced corrections to charged-Higgs and chargino interactions have been discussed for MFV scenarios with large $\tan\beta$ in Refs. [48, 60]; in this section we derive the corresponding results for the non-MFV case using our diagrammatic approach.

Flavour-changing self-energies lead to antihermitian corrections to the matrices $U_L^{(0)q}$. Charged-Higgs and chargino couplings also involve right-handed fields; the corresponding corrections to $U_R^{(0)q}$ are obtained by simply exchanging the chiralities in the expressions for

ΔU_L^q (cf. Eqs. (3.20) and (3.29)). The CKM matrix which enters charged-Higgs or chargino vertices is not the physical one, because in these cases $V^{(0)}$ does not add up to V together with enhanced loop corrections. The charged Higgs interaction $\bar{u}_f H^+ d_i$ has the Feynman rule

$$-i\Lambda_{H^+}^{(0)} = i\left(Y^{u_f *}V_{fi}^{(0)}\cos\beta P_L + V_{fi}^{(0)}Y^{d_i}\sin\beta P_R\right). \tag{3.30}$$

The effect of self-energies in the external legs is included by substituting this Feynman rule with

$$\Lambda_{H^+}^{(0)} \longrightarrow -\sum_{j,k=1}^{3}\left[\left(1+\Delta U_R^{u\dagger}\right)_{fj}Y^{u_j *}\left(1-\Delta U_L^{u\dagger}\right)_{jk}V_{ki}\cos\beta P_L\right.$$
$$\left. + V_{fj}\left(1-\Delta U_L^{d}\right)_{jk}Y^{d_k}\left(1+\Delta U_R^{d}\right)_{ki}\sin\beta P_R\right]. \tag{3.31}$$

Using the explicit expression for $\Delta U_{L,R}^q$ given in (3.20) and expressing Y^{q_j} in terms of quark masses through (3.28) the substitution rule of (3.31) becomes

$$\Lambda_{H^+}^{(0)} \to \tag{3.32}$$
$$-\sum_{j=1}^{3}\left[\begin{pmatrix}\frac{m_{u_1}}{1+\Delta_{u_1}} & \frac{-\Sigma_{12}^{uRL}}{1+\Delta_{u_2}} & \frac{-\Sigma_{13}^{uRL}}{1+\Delta_{u_3}} \\ \frac{-\Sigma_{21}^{uRL}}{1+\Delta_{u_2}} & \frac{m_{u_2}}{1+\Delta_{u_2}} & \frac{-\Sigma_{23}^{uRL}}{1+\Delta_{u_3}} \\ \frac{-\Sigma_{31}^{uRL}}{1+\Delta_{u_3}} & \frac{-\Sigma_{32}^{uRL}}{1+\Delta_{u_3}} & \frac{m_{u_3}}{1+\Delta_{u_3}}\end{pmatrix}_{fj}\frac{V_{ji}\cos\beta}{v_u}P_L + \frac{V_{fj}\sin\beta}{v_d}\begin{pmatrix}\frac{m_{d_1}}{1+\Delta_{d_1}} & \frac{-\Sigma_{12}^{dLR}}{1+\Delta_{d_2}} & \frac{-\Sigma_{13}^{dLR}}{1+\Delta_{d_3}} \\ \frac{-\Sigma_{21}^{dLR}}{1+\Delta_{d_2}} & \frac{m_{d_2}}{1+\Delta_{d_2}} & \frac{-\Sigma_{23}^{dLR}}{1+\Delta_{d_3}} \\ \frac{-\Sigma_{31}^{dLR}}{1+\Delta_{d_3}} & \frac{-\Sigma_{32}^{dLR}}{1+\Delta_{d_3}} & \frac{m_{d_3}}{1+\Delta_{d_3}}\end{pmatrix}_{ji}P_R\right]$$

We observe a cancellation between the inverse quark masses in ΔU_L^q (see (3.20)) and the factors of m_{q_i} from the Y^{q_i}'s in the effective off-diagonal couplings.

For all Higgs processes the genuine vertex correction $\Lambda_{H^+}^{(1)}$ is of the same order as the diagrams with self energies in the external leg. Furthermore, in the absence of terms with the "wrong" vev in the squark mass matrices there is an exact cancellation between the genuine vertex correction and the external self-energies in the decoupling limit. This cancellation was observed for neutral Higgs couplings in Ref. [61] and can be understood from Fig. 3.3: The upper right diagram involving A_{fi}^d merely renormalizes the Yukawa coupling and maintains the type-II 2HDM structure of the tree–level Higgs sector. Therefore the loop-corrected Higgs couplings are identical to the tree-level ones, provided they are expressed in terms of V_{fi} and the physical quark masses. In our diagrammatic approach A_{fi}^q enters both the proper vertex correction and Σ_{jk}^{qLR} and cancels from the combined result. We neglect all external momenta, so that our expression for $\Lambda_{H^+}^{(1)}$ is not valid for top or H^+ decays unless the gluino or the squarks appearing in the loop function are much heavier

3.4 Renormalization of the CKM matrix

than the top quark and the charged Higgs boson. The proper vertex correction, to be added to Eqs. (3.31) and (3.32), reads:

$$
\begin{aligned}
\Lambda_{H^+}^{(1)} = -\frac{2\alpha_s}{3\pi} m_{\tilde{g}} \sum_{s,t=1}^{6} \sum_{k,l=1}^{3} & \left\{ \left(V_{s\,fk}^{(0)\,u\,LL} V_{t\,li}^{(0)\,d\,RR} P_R + V_{s\,fk}^{(0)\,u\,RL} V_{t\,li}^{(0)\,d\,RL} P_L \right) H_{kl}^{+\,LR} \right. \\
& + \left(V_{s\,fk}^{(0)\,u\,LR} V_{t\,li}^{(0)\,d\,LR} P_R + V_{s\,fk}^{(0)\,u\,RR} V_{t\,li}^{(0)\,d\,LL} P_L \right) H_{kl}^{+\,RL} \\
& + \left(V_{s\,fk}^{(0)\,u\,LL} V_{t\,li}^{(0)\,d\,LR} P_R + V_{s\,fk}^{(0)\,u\,RL} V_{t\,li}^{(0)\,d\,LL} P_L \right) H_{kl}^{+\,LL} \\
& \left. + \left(V_{s\,fk}^{(0)\,u\,LR} V_{t\,li}^{(0)\,d\,RR} P_R + V_{s\,fk}^{(0)\,u\,RR} V_{t\,li}^{(0)\,d\,RL} P_L \right) H_{kl}^{+\,RR} \right\} \\
& \times C_0 \left(m_{\tilde{u}_s}, m_{\tilde{d}_t}, m_{\tilde{g}} \right)
\end{aligned}
$$
(3.33)

The coefficients $H_{kl}^{+\,AB}$ are given in (10.6) of the appendix.

In the case of chargino interactions we must take into account that a squark never comes with an enhanced self-energy, even if the squark line is an external line of the considered Feynman diagram. The Feynman rules for the chargino-quark-squark coupling are given in equation (10.4) and (10.5). Again we include the self-energy corrections, and express $V^{(0)}$ in terms of the physical CKM matrix. Then the effective chargino couplings read:

$$
\begin{aligned}
\Gamma_{u_i \tilde{d}_s}^{\tilde{\chi}_k^{\pm}\,L*\,\text{eff}} &= \sum_{j,k=1}^{3} U_{k2}^{\tilde{\chi}^{\pm}} W_{j+3,t}^{\tilde{d}*} Y^{d_j} \left(1 - \Delta U_L^{d\dagger}\right)_{jk} V_{ki}^* - g_w U_{k1}^{\tilde{\chi}^{\pm}} \sum_{j,k=1}^{3} W_{jt}^{\tilde{d}*} \left(1 - \Delta U_L^{d\dagger}\right)_{jk} V_{ki}^*, \\
\Gamma_{u_i \tilde{d}_s}^{\tilde{\chi}_k^+\,R*\,\text{eff}} &= V_{k2}^{\tilde{\chi}^{\pm}*} \sum_{j,k,l,m=1}^{3} W_{jt}^{\tilde{d}*} \left(1 - \Delta U_L^{d\dagger}\right)_{jk} V_{kl}^* (1 - \Delta U_L^u)_{lm} Y^{u_m} \left(1 + \Delta U_R^u\right)_{mi}, \\
\Gamma_{d_i \tilde{u}_s}^{\tilde{\chi}_k^{\pm}\,L*\,\text{eff}} &= \sum_{j,k=1}^{3} V_{k2}^{\tilde{\chi}^{\pm}*} W_{j+3,t}^{\tilde{u}*} Y^{u_j} \left(1 - \Delta U_L^{u\dagger}\right)_{jk} V_{ki} - g_w V_{k1}^{\tilde{\chi}^{\pm}*} \sum_{j,k=1}^{3} W_{jt}^{\tilde{u}*} \left(1 - \Delta U_L^{u\dagger}\right)_{jk} V_{ki}, \\
\Gamma_{d_i \tilde{u}_s}^{\tilde{\chi}_k^{\pm}\,R*\,\text{eff}} &= U_{k2}^{\tilde{\chi}^{\pm}} \sum_{j,k,l,m=1}^{3} W_{jt}^{\tilde{u}*} \left(1 - \Delta U_L^{u\dagger}\right)_{jk} V_{kl} (1 - \Delta U_L^d)_{lm} Y^{d_m} \left(1 + \Delta U_R^d\right)_{mi}.
\end{aligned}
$$
(3.34)

We have seen in this section that in the case of non-minimal flavour violation the CKM matrix (including loop corrections) entering charged Higgs and quark-squark-chargino vertices is not simply the physical one. Instead it has to be corrected according to (3.31) or (3.32) and (3.34), leading to potentially large effects.

4. Naturalness constraints

In this chapter we are going to give a complete evaluation of the all possible constraints on the SUSY breaking sector from 't Hooft's naturalness argument.

Large accidental cancellations between the SM and supersymmetric contributions are, as already mentioned in the introduction, unlikely and from the theoretical point of view undesirable. Requiring the absence of such cancellations is a commonly used fine-tuning argument, which is also employed in standard FCNC analyses of the $\delta_{ij}^{q\,XY}$'s [9–14, 25]. Furthermore, in our case we can use 't Hooft's naturalness argument since we gain a flavour symmetry if the light fermion masses are generated radiatively. Therefore, the situation is different from e.g. the little hierarchy problem, where no additional symmetry is involved. First of all, it is important to note that all off-diagonal elements of the fermion mass matrices have to be smaller than the average of their assigned diagonal elements

$$\left(\Delta m_F^2\right)_{XY}^{ij} < \sqrt{m_{\tilde{f}_{iX}}^2 m_{\tilde{f}_{jY}}^2}, \tag{4.1}$$

since otherwise one sfermion mass eigenvalue is negative. We note that in Ref. [10] this constraint is not imposed.

All constraints in this section are non-decoupling since we compute corrections to the Higgs-quark-quark coupling which are of dimension 4. Therefore, our constraints on the soft-supersymmetry-breaking parameters do not vanish in the limit of infinitely heavy SUSY masses but rather converge to a constant. However, even though $\delta_{ij}^{f\,LR}$ is a dimensionless parameter it does not only involve SUSY parameter. It is also proportional to a vacuum expectation and therefore scales like v/M_{SUSY}. Thus, our constraints on $\delta_{ij}^{f\,LR}$ do not approach a constant for $M_{\text{SUSY}} \to \infty$ but rather get stronger. Similar effects occur in Higgs-mediated FCNC processes which decouple like $1/M_{\text{Higgs}}^2$ rather than $1/M_{\text{SUSY}}^2$ [49, 56, 62]. However, Higgs-mediated effects can only be induced within supersymmetry in the presence of non-holomorphic terms which are not required for our constraints. An example of a non-decoupling Higgs-mediated FCNC process is the observable $R_K = \Gamma\left(K \to e\nu\right)/\Gamma\left(K \to \mu\nu\right)$ that is currently analyzed by the NA62-experiment.

In this case Higgs contributions can induce deviations from lepton flavour universality [63–65].

4.1. Constraints on flavor-diagonal mass insertions at one loop

The diagonal elements of the A-terms can be constrained from the fermion masses by demanding that $\Sigma_{ii}^{f\,LR\,(1)} \leq m_{f_i}$ (see (3.19)). The bounds on the flavor-conserving A-term for the up, charm, down and strange quarks as well as for the electron and muon are shown in Fig. (4.1). The upper bound derived from the fermion mass is roughly given by

$$\left|\delta_{ii}^{q\,LR}\right| \lesssim \frac{3\pi\, m_{q_i}(M_{SUSY})}{\alpha_s(M_{SUSY}) M_{SUSY}} \tag{4.2}$$

for quarks and

$$\left|\delta_{ii}^{\ell\,LR}\right| \lesssim \frac{8\pi m_{\ell_i}}{\alpha_1 M_{SUSY}} \tag{4.3}$$

for leptons in the case of equal SUSY masses. In the lepton case (4.3) can be further simplified, since we can neglect the running of the masses:

$$\begin{aligned}\left|\delta_{11}^{\ell\,LR}\right| &\lesssim 0.0025 \left(\tfrac{500\,\text{GeV}}{M_{\text{SUSY}}}\right), \\ \left|\delta_{22}^{\ell\,LR}\right| &\lesssim 0.5 \left(\tfrac{500\,\text{GeV}}{M_{\text{SUSY}}}\right).\end{aligned} \tag{4.4}$$

However, as already pointed out in Ref. [66] a muon mass that is solely generated radiatively potentially leads to measurable contributions to the muon anomalous magnetic moment. This arises from the same one-loop diagram as $\Sigma_{22}^{\ell\,LR}$ with an external photon attached. Therefore, the SUSY contribution is not suppressed by a loop factor compared to the case with tree-level Yukawa couplings.

4.2. Constraints on flavor-off-diagonal mass insertions from CKM and PMNS renormalization

4.2.1. CKM matrix

We use the standard parameterisation for the CKM matrix and quark masses defined in the $\overline{\text{MS}}$ scheme with the central PDT values [67]. As discussed at the end of Sect. 3.4, the masses enter the loop contributions to V in (3.20) at the renormalisation scale $Q = M_{\text{SUSY}}$, at which the self-energies are calculated.

Because of the V–A structure of the W vertex only self-energies Σ_{fi}^{qLR} with $f < i$ enter the renormalisation of the CKM matrix in the approximation $m_{q_f} = 0$ (see (3.20)). This implies that for $f < i$ only δ_{fi}^{qLR} is constrained in the leading order of the mass insertion approximation. In order to constrain δ_{fi}^{qLL} or δ_{fi}^{qRR} an additional insertion of some δ_{jk}^{qLR} is needed. Even three insertions are needed to involve δ_{fi}^{qRL}. These diagrams with multiple insertions of δ_{jk}^{qXY} can give useful bounds on one of these quantities only if the δ_{jk}^{qLR} providing the needed chirality-flip is large. Indeed, we find useful bounds on $|\delta_{13}^{dLL}|$ in the large-$\tan\beta$ scenario, where $\delta_{33}^{dLR}/m_{d_3}$ is large. The analogous contributions involving δ_{ij}^{dRR} are suppressed with respect to those with $\delta_{ij}^{dLL}\delta_{jj}^{dLR}$ by a small ratio of quark masses. The upper limits on δ_{fi}^{qRR} and δ_{fi}^{qRL} from vacuum stability [68], electric dipole moments and FCNC processes are stronger than ours.

We use the following input parameters [67]:

$$\begin{aligned}
\overline{m}_s(2\text{ GeV}) &= 0.095\text{ GeV}, & \overline{m}_c(\overline{m}_c) &= 1.25\text{ GeV}, \\
\overline{m}_b(\overline{m}_b) &= 4.2\text{ GeV}, & \overline{m}_t(\overline{m}_t) &= 166\text{ GeV}, \\
|V_{us}| &= 0.227, & |V_{ub}| &= 0.00396, & |V_{cb}| &= 0.0422.
\end{aligned} \quad (4.5)$$

4.2.2. Down-sector

We present our bounds on $|\delta_{ij}^{dLR}|$ and $|\delta_{ij}^{dLL}|$ in Sects. 4.2.2 and 4.2.2, respectively.

Constraints on $|\delta_{ij}^{dLR}|$

Constraints from $\mathbf{V_{us}, V_{cd}}$: V_{us} and V_{cd} are experimentally well known. Their absolute values are nearly equal and they have opposite sign in the standard parameterisation, which is respected by the corrections (3.20). Fig. 4.2 shows the dependence of the constraints on the squark mass with different ratios $m_{\tilde{g}}/m_{\tilde{q}}$. In the approximation $m_{d_1} = 0$, only δ_{12}^{dLR} is constrained. Looking at the dependence on the gluino mass (Fig. 4.3), it is interesting to note that there is a minimum for $m_{\tilde{g}} \approx 1.5 m_{\tilde{q}}$.

Constraints from $\mathbf{V_{cb}, V_{ts}}$: In this case, the situation is nearly the same as in the case of V_{us}, except that the constraints are weaker (see Fig. 4.4), because m_b is much larger than m_s. $|V_{ts}|$ is essentially fixed by the measured value of $|V_{cb}|$ through CKM unitarity.

Constraints from $\mathbf{V_{ub}}$: The last pair of of CKM elements to be discussed is (V_{ub}, V_{td}). In this case $|V_{ub}|$ does not fix $|V_{td}|$, because $|V_{td}|$ is largely affected by the CKM phase. Now

4.2 Constraints on flavor-off-diagonal mass insertions from CKM and PMNS renormalization

$|V_{ub}|$ is experimentally better known than $|V_{td}|$, because V_{td} is extracted from FCNC loop processes by comparing the corresponding experimental result with the SM prediction. In beyond-SM scenarios, this is not a valid procedure anymore, because the new particles will alter the FCNC loop processes. Therefore, we can only exploit the constraint from $|V_{ub}|$. The anti-hermiticity of ΔU_L^d in (3.20) implies that $\delta_{13}^{d\,LR}$ gives a negative contribution of the same size to V_{td}. This cannot be the whole contribution, since $|V_{ub}| \neq |V_{td}|$ and in the standard CKM parameterisation $\mathrm{Re}\,V_{td}$ and $\mathrm{Re}\,V_{ub}$ are both positive. The hierarchical structure of the CKM matrix is responsible for a second contribution: Since $V_{ub} \propto \lambda$ and $V_{cb} \propto \lambda^2$ is important as well. This diagram corresponds to the correction $\Delta U_L^{q\,(2)}{}_{31}$ given in equation (3.23) which adds in this case a contribution of $V_{us} * V_{cb} = 0.0088$ to V_{td}. Together with the one-loop contribution from $\delta_{13}^{d\,LR}$, this yields the correct value for V_{td}. We stress that this does not imply any additional constraint on the SUSY parameters entering the self-energies, to order λ^3 we just reproduced a unitarity relation of the CKM elements: $V_{td} = -V_{ub}^* + V_{cd}V_{ts} \simeq -V_{ub}^* - V_{us}^*V_{ts}$, which (with insertion of $V_{tb} \simeq V_{ud} \simeq 1$) equals the product of the first and third rows of the CKM matrix.

The last possible constraint is the phase of the CKM matrix, which one could infer from $\gamma = \arg(-V_{ub}^* V_{ud}/(V_{cb}^* V_{cd}))$. However, since γ is large, no fine-tuning argument can be applied to derive bounds. Only in a given scenario of radiatively generated CKM elements, the measured value of γ can be used to derive a constraint on the complex phases in the mass matrix of (2.17).

Constraints on $\delta_{ij}^{d\,LL}$

In the presence of large chirality-flipping flavour-diagonal elements in the squark mass matrix, also $\delta_{ij}^{q\,LL}$ can be constrained. This is the case for large A_{jj}^q terms or (if $q = d$) for a large value of $\mu \tan\beta$. Here we only consider the second possibility, which is widely studied in the literature. The strongest constraints are obtained for $\delta_{13}^{d\,LL}$, because V_{ub} is the smallest entry of the CKM matrix. We have included the correction term Δ_b of (3.28) in our analysis. Our result is shown in Fig. 4.6. Our constraint is compatible with the experimental bound on $Br(B_d \to \mu^+\mu^-)$ for values of $\tan\beta$ around 30 or below [69].

We next discuss the constraint on $\delta_{23}^{d\,LL}$: It is clear that our bound will be looser by a factor of $|V_{cb}/V_{ub}|$. Furthermore, for large $\tan\beta$ and typical values of the massive SUSY parameters we find $Br(B_s \to \mu^+\mu^-)$ more constraining. To find bounds on $\delta_{23}^{d\,LL}$ from $|V_{cb}|$ which comply with $Br(B_s \to \mu^+\mu^-)$ we need a smaller value of $\tan\beta$ around 20 and therefore a quite large value for μ, if the masses of the non-standard Higgs bosons are around 500 GeV. We do not include a bound on $\delta_{23}^{d\,LL}$ in our table of results in Sect. 4.2.4.

4.2.3. Up-sector

In the up-sector, everything is in straight analogy to the down-sector. The only difference is that the constraints are weaker because of the larger charm and top masses. But the upper bounds are still restrictive, except the ones obtained from V_{ts} or V_{cb} (see Fig. 4.8). Remarkably, we now have a powerful constraint on $\delta_{13}^{u\,LR}$ from the second diagram in Fig. 3.5 with $q_{u_f} = u$ and $q_{d_i} = b$.

4.2.4. Comparison with previous bounds

In this section we compare our bounds with those in the literature, derived from FCNC processes [11, 14 25 70] and vacuum stability (VS) bounds [68]. We take $M_{\text{SUSY}} = \sqrt{\left[M_{\tilde{q}}^2\right]_{ss}} = m_{\tilde{g}} = 1000$ GeV:

quantity	our bound	bound from FCNC's		bound from VS [68]
$|\delta_{12}^{d\,LR}|$	≤ 0.0011	≤ 0.006	K mixing [11]	$\leq 1.5 \times 10^{-4}$
$|\delta_{13}^{d\,LR}|$	≤ 0.0010	≤ 0.15	B_d mixing [14]	≤ 0.05
$|\delta_{23}^{d\,LR}|$	≤ 0.010	≤ 0.06	$B \to X_s\gamma; X_s l^+l^-$ [70]	≤ 0.05
$|\delta_{13}^{d\,LL}|$	≤ 0.032	≤ 0.5	B_d mixing [14]	—
$|\delta_{12}^{u\,LR}|$	≤ 0.011	≤ 0.016	D mixing [25]	$\leq 1.2 \times 10^{-3}$
$|\delta_{13}^{u\,LR}|$	≤ 0.062	—		≤ 0.22
$|\delta_{23}^{u\,LR}|$	≤ 0.59	—		≤ 0.22

Our value for $\delta_{13}^{d\,LL}$ is calculated with $\dfrac{\mu \tan \beta}{1 + \Delta_b} = 20$ TeV. The quoted bound on $\delta_{23}^{d\,LR}$ from $b \to s\gamma$ and $B \to X_s l^+l^-$ has been rescaled by an approximate factor of 3 from the value quoted for $M_{\text{SUSY}} = 350$ GeV in Ref. [70]. The VS bounds on $\delta_{ij}^{u\,LR}$ have also been obtained by scaling the quoted values for $M_{\text{SUSY}} = 500$ GeV of Ref. [68] by a factor of $1/2$. The VS bounds on $\delta_{13}^{u\,LR}$ and $\delta_{23}^{u\,LR}$ are obtained by multiplying the bound on $\delta_{12}^{u\,LR}$ with m_t/m_c. FCNC effects are decoupling and scale as $1/M_{\text{SUSY}}^2$, but the constraints on $\delta_{ij}^{q\,LR}$ are proportional to M_{SUSY} rather than M_{SUSY}^2, because the definition of $\delta_{ij}^{q\,LR}$ involves a factor of v/M_SUSY. Both our constraints and the VS bounds on the trilinear SUSY-breaking terms are independent of M_SUSY (i.e. non-decoupling), so that the bounds on $\delta^{q\,LR}_{ij}$ scale like $1/M_\text{SUSY}$. We conclude that all our bounds on $\delta^{d\,LR}_{ij}$ are more restrictive than those from FCNC processes for $M_\text{SUSY} \geq 500$ GeV, and our bound on $\delta^{u\,LR}_{12}$ is stronger than the quoted FCNC bound for $M_\text{SUSY} \geq 900$ GeV.

Substantially stronger bounds than ours are only listed for the VS bounds on $|\delta^{u\,LR}_{23}|$, $|\delta^{u\,LR}_{12}|$ and $|\delta^{d\,LR}_{12}|$. However, the VS bounds related to the latter two quantities are of the form

$$A^q_{12} < Y^{q_2} f, \tag{4.6}$$

where $f = \mathcal{O}(M_\text{SUSY})$ depends on other massive parameters of the scalar potential. The bounds are obtained by studying the scalar potential at tree level and Y^{q_2} enters the analysis through the quartic coupling of strange squarks to Higgs bosons. The smallness of Y^{q_2} makes this coupling sensitive to large loop corrections and the quoted bounds have to be considered as rough estimates at best. Our results for $\delta^{q\,LR}_{12}$ rest on a firmer footing.

4.2.5. Threshold corrections to PMNS matrix

Up to now, we have only an upper bound for the matrix element $U_{e3} = \sin\theta_{13} e^{-i\delta}$ and thus for the mixing angle θ_{13}; the best-fit value is at or close to zero: $\theta_{13} = 0.0^{+7.9}_{-0.0}$ [71]. It might well be that it vanishes at tree level due to a particular symmetry and obtains a non-zero value due to corrections. So we can ask the question if threshold corrections to the PMNS matrix could spoil the prediction $\theta_{13} = 0°$ at the weak scale. We demand the absence of fine-tuning for these corrections and therefore require that the SUSY loop contributions do not exceed the value of U_{e3},

$$|\Delta U_{e3}| \leq \left|U^\text{phys}_{e3}\right|. \tag{4.7}$$

The renormalization of the PMNS matrix is described in detail in [72], where the on-shell scheme was used. As discussed in Sec. (3.1) we also use the $\overline{\text{MS}}$ scheme in this section. Then the physical PMNS matrix is given by:

$$U^\text{phys} = U^{(0)} + \Delta U, \tag{4.8}$$

where ΔU should not be confused with the wave function renormalization $\Delta U^{f\,L}$. Then ΔU is given by

$$\Delta U = \left(\Delta U^{l\,L}\right)^T U^{(0)}. \tag{4.9}$$

Note that in contrast to the corrections to the CKM matrix, there is a transpose in ΔU^{lL}, because the first index of the PMNS matrix corresponds to down-type fermions and not to up-type fermion as in the CKM matrix. Only the corrections to the small element U_{e3} can be sizeable, since all other elements are of order one. If we set all off-diagonal element to zero except for $\delta_{LR}^{13} \neq 0$, we get

$$\Delta U_{e3} = \frac{\Delta U_{31}^{lL} U_{\tau 3}^{\text{phys}} - U_{e3}^{\text{phys}} \left|\Delta U_{31}^{lL}\right|^2}{1 + \left|\Delta U_{11}^{lL}\right|^2} \quad (4.10)$$

$$\approx -U_{\tau 3}^{\text{phys}} \frac{\Sigma_{31}^{lRL}}{m_\tau}.$$

Note that here, in contrast to the renormalization of the CKM matrix, the physical PMNS element appears. This is due to the fact that one has to solve the linear system in (4.9) as described in [72]. By means of the fine-tuning argument we can in principle derive upper bounds for δ_{13}^{lLR}. The results depend on the SUSY mass scale M_{SUSY} and the assumed value for θ_{13}.

Here, we consider the corrections stemming from flavor-violating A-terms to the small matrix element U_{e3}. The δ_{13}^{lLL}-contribution was already studied in [72] with the result that they are negligible small. We also made a comment about the δ_{13}^{lLR}-contribution which is outlined in more detail. Our results depend on the overall SUSY mass scale, the value of θ_{13} and of δ_{13}^{lLR}. In Fig. (4.9) you can see the percentage deviation of U_{e3} through this SUSY loop corrections in dependence of δ_{13}^{lLR} (top) and θ_{13} (bottom) for $M_{\text{SUSY}} = 1000$ GeV. The constraints on δ_{13}^{lLR} get stronger with smaller θ_{13} and with larger M_{SUSY}. In Fig. (4.10) the excluded $\left(\theta_{13}, \delta_{13}^{lLR}\right)$-region is below the curves for different M_{SUSY} scales. The derived bound can be simplified to

$$\left|\delta_{13}^{lLR}\right| \lesssim 0.2 \left(\frac{500 \text{ GeV}}{M_{\text{SUSY}}}\right) |\theta_{13} \text{ in degrees}|. \quad (4.11)$$

Exemplarily, we get for reasonable SUSY masses of $M_{\text{SUSY}} = 1000$ GeV and $\theta_{13} = 3°$ an upper bound of $\left|\delta_{13}^{lLR}\right| \leq 0.3$. The constraints on δ_{13}^{lLR} from $\tau \to e\gamma$ are of the order of 0.02 [72] and in general better than our derived bounds if θ_{13} is non-zero. As an important consequence, we note that $\tau \to e\gamma$ impedes any measurable correction from supersymmetric loops to U_{e3}: E.g. for sparticle masses of 500 GeV we find $|\Delta U_{e3}| \leq 10^{-3}$ corresponding to a correction to the mixing angle θ_{13} of at most $0.06°$. That is, if the DOUBLE CHOOZ experiment measures $U_{e3} \neq 0$, one will not be able to ascribe this result to the SUSY breaking sector. Stated positively, $U_{e3} \gtrsim 10^{-3}$ will imply that at low energies the flavor symmetries imposed on the Yukawa sector to motivate tri-bimaximal mixing are violated. This finding confirms the pattern found in [72] where the product $\delta_{13}^{l\,LL}\delta_{33}^{l\,LR}$ has been studied instead of $\delta_{13}^{l\,LR}$.

4.3. Constraints from two-loop corrections to fermion masses

Combining two flavor-violating self-energies can have sizable impacts on the light fermion masses according to (3.22). Requiring that no large numerical cancellations should occur between the tree-level mass (which is absent in the case of a radiative fermion mass) and the supersymmetric loop corrections we can derive bounds on the products $\delta_{ij}^{f\,LR}\delta_{ji}^{f\,LR}$ which contain the so far less constrained elements $\delta_{ij}^{f\,LR}$, $i > j$.

We apply the fine-tuning argument to the two-loop contribution originating from flavor-violating A-terms, e.g. $\left|\Sigma_{11}^{f\,LR(2)}\right| \leq m_{f_1}$. The bound $\Sigma_{11}^{f\,LR(2)} = m_{f_1}$ corresponds to a 100% change in the fermion mass through supersymmetric loop corrections which is equivalent to the case that the fermion Yukawa coupling vanishes. The upper bound depends on the overall SUSY mass scale and is roughly given as

$$\left|\delta_{i3}^{q\,LR}\delta_{3i}^{q\,LR}\right| \lesssim \frac{9\pi^2 m_{q_i} m_{q_3}(M_{SUSY})}{(\alpha_s(M_{SUSY})M_{SUSY})^2}, \quad i \neq 3 \quad (4.12)$$

for quarks and

$$\left|\delta_{13}^{l\,LR}\delta_{31}^{l\,LR}\right| \lesssim \frac{64\pi^2 m_{l_1} m_{l_3}}{(\alpha_1 M_{SUSY})^2} \quad (4.13)$$

for leptons. Again, (4.13) can be further simplified

$$\left|\delta_{13}^{l\,LR}\delta_{31}^{l\,LR}\right| \leq 0.021 \left(\frac{500\,\text{GeV}}{M_{SUSY}}\right)^2. \quad (4.14)$$

It is important to calculate bounds on the product $\delta_{13}^{u\,LR}\delta_{31}^{u\,LR}$ because $\delta_{31}^{u\,LR}$ cannot be severely constrained by FCNC processes [73]. As studied in Ref. [74], single-top production involves the same mass insertion $\delta_{31}^{u\,LR}$ which can also induce a right-handed W coupling if at the same time $\delta_{33}^{d\,LR} \neq 0$ (see chapter 8). Therefore our bound can be used to place a constraint on this cross section. Also the product $\delta_{23}^{u,l\,LR}\delta_{32}^{u,l\,LR}$ cannot be constrained, since the muon and the charm are too heavy. However, $\delta_{23}^{d\,LR}\delta_{32}^{d\,LR}$ can be constrained and the results are depicted in Fig. (4.11). In the quark case also the bounds from the CKM renormalization on $\delta_{13,23}^{q\,LR}$ are taken into account.

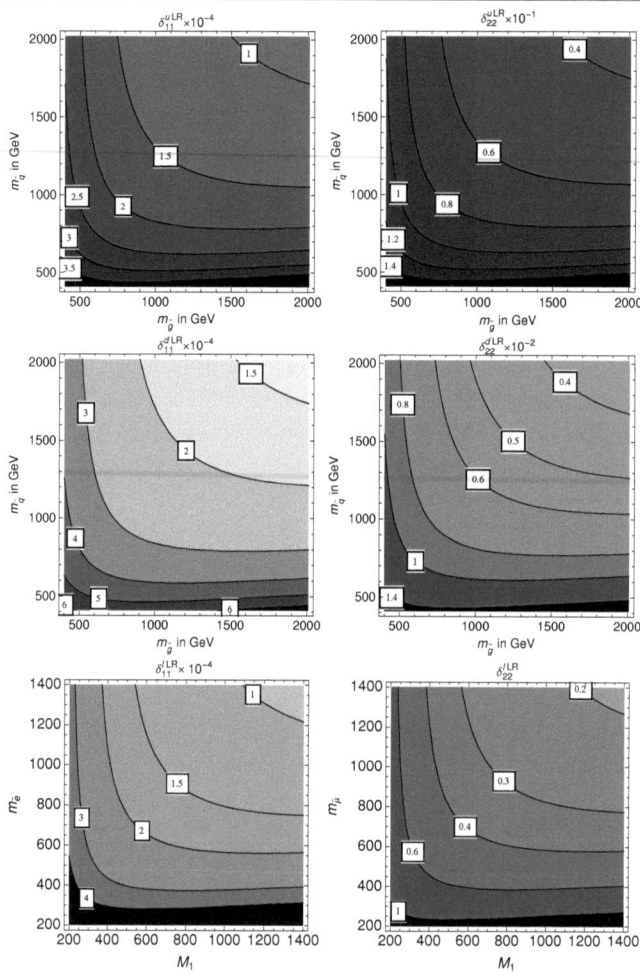

Figure 4.1: Constraints on the diagonal mass insertions $\delta_{11,22}^{\ell,u,dLR}$ obtained by applying 't Hooft's naturalness criterion.

4.3 Constraints from two-loop corrections to fermion masses

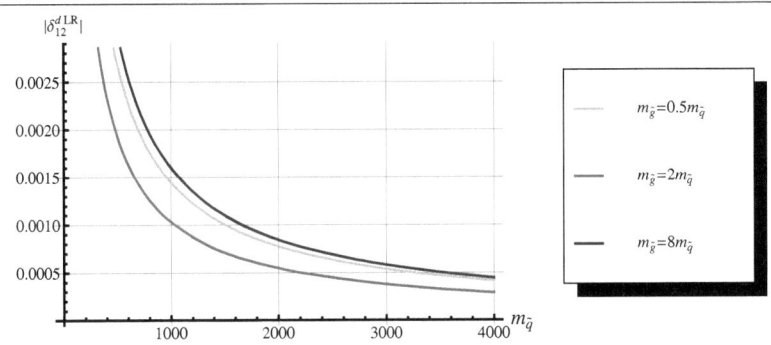

Figure 4.2: Constraints on $|\delta_{12}^{d\,LR}|$ from $|V_{us}|$ or $|V_{cd}|$ as a function of the squark mass. The constraints become stronger with growing M_{SUSY}, because $\delta_{ij}^{q\,LR} \propto \dfrac{v}{M_{\text{SUSY}}}$.

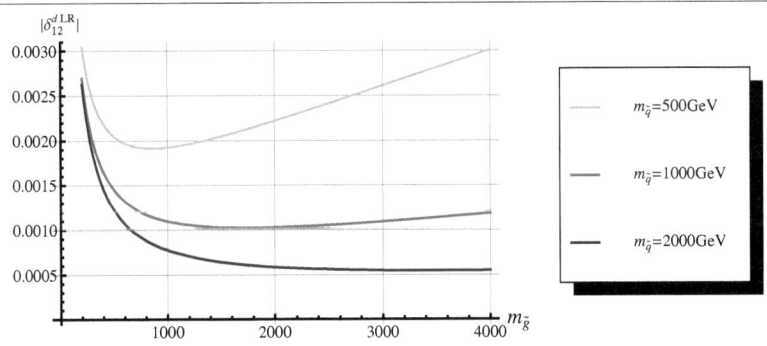

Figure 4.3: Constraints on $|\delta_{12}^{d\,LR}|$ from $|V_{us}|$ (or $|V_{cd}|$) as a function of the gluino mass.

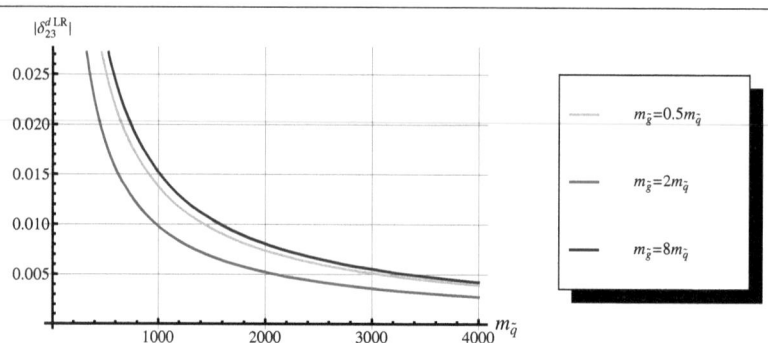

Figure 4.4: Constraints on $|\delta_{23}^{d\,LR}|$ from $|V_{cb}|$ (or $|V_{ts}|$) as a function of the squark mass.

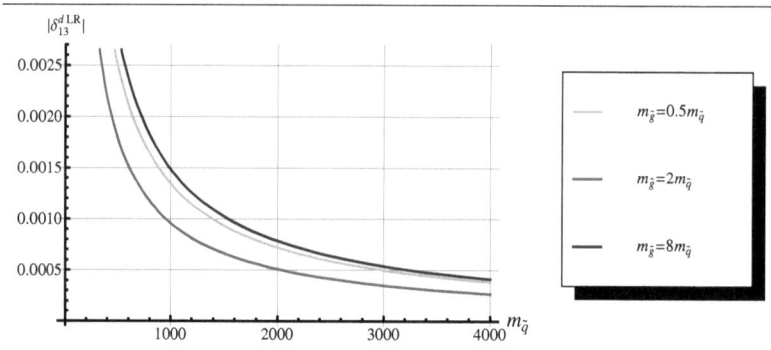

Figure 4.5: Constraints on $|\delta_{13}^{d\,LR}|$ from $|V_{ub}|$ as a function of the squark mass.

4.3 Constraints from two-loop corrections to fermion masses

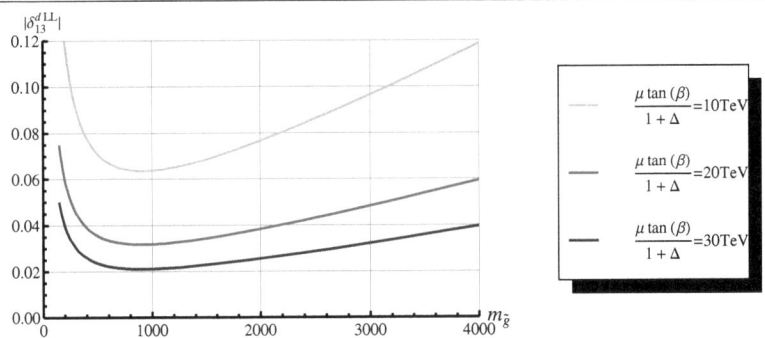

Figure 4.6: Constraints on $|\delta_{13}^{d\,LL}|$ from $|V_{ub}|$ as a function of the gluino mass for different values of $\mu \tan\beta/(1+\Delta_b)$.

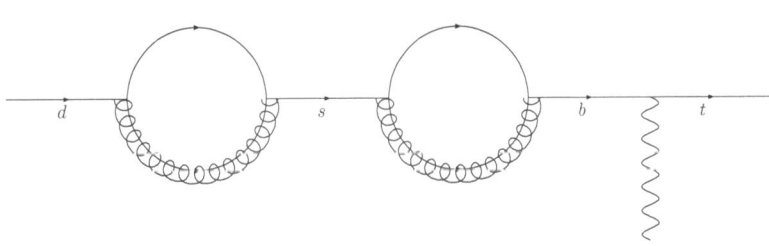

Figure 4.7: Numerically important two-loop correction to V_{td}. The analogous diagram also exists in the up sector.

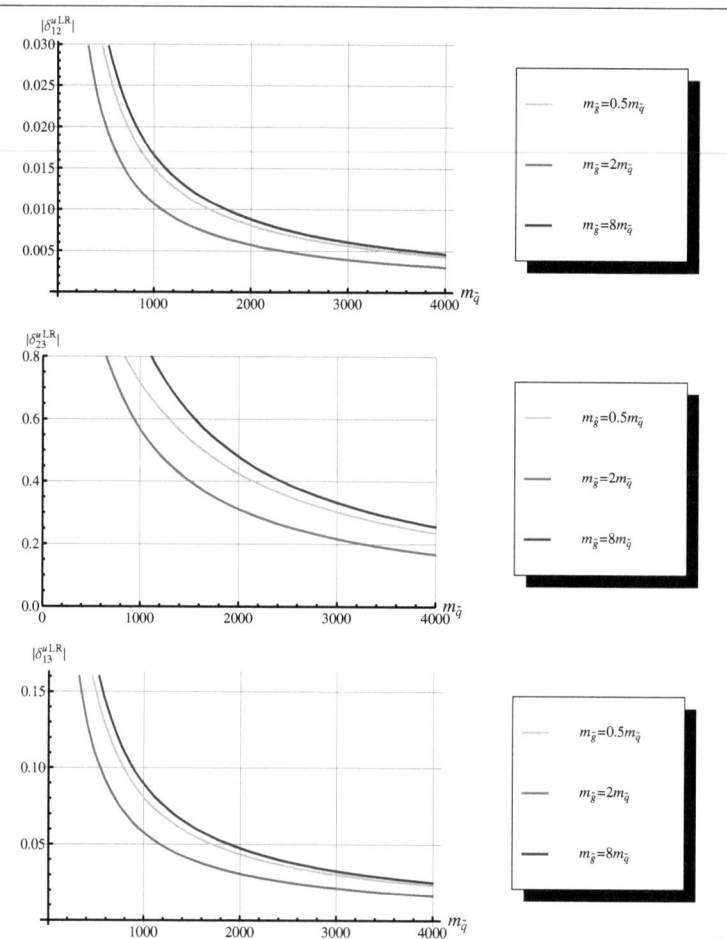

Figure 4.8: Constraints on $\delta_{12}^{u\,LR}$, $\delta_{23}^{u\,LR}$ and $\delta_{13}^{u\,LR}$ as a function of the squark mass for different ratios of $m_{\tilde{g}}/m_{\tilde{q}}$.

4.3 Constraints from two-loop corrections to fermion masses

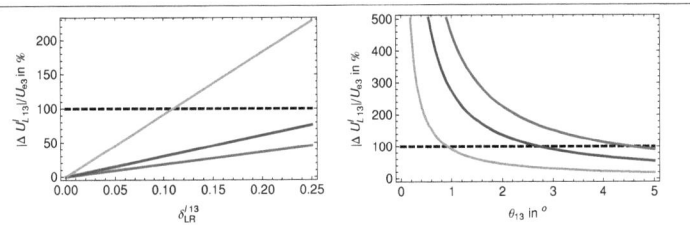

Figure 4.9: $|\Delta U_{e3}|/U_{e3}$ in %. Top: as a function of $\delta_{13}^{l\,LR}$ for $M_{\text{SUSY}} = 1000$ GeV and different values of θ_{13} (green 1°; blue: 3°; red: 5°). Bottom: as a function of θ_{13} for $M_{SUSY} = 1000$ GeV and different values of $\delta_{LR}^{l\,13}$ (red: $\delta_{13}^{l\,LR} = 0.5$; blue: $\delta_{13}^{l\,LR} = 0.3$; green: $\delta_{13}^{l\,LR} = 0.1$) (both from top to bottom)

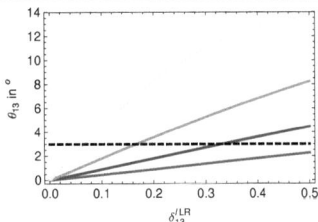

Figure 4.10: The excluded $(\theta_{13}, \delta_{13}^{l\,LR})$-region is below the curves for (from bottom to top) $M_{\text{SUSY}} = 500$ GeV (red), 1000 GeV (blue), 2000 GeV (green) and 5000 GeV (yellow). The black dashed line denotes the future experimental sensitivity to $\theta_{13} = 3°$.

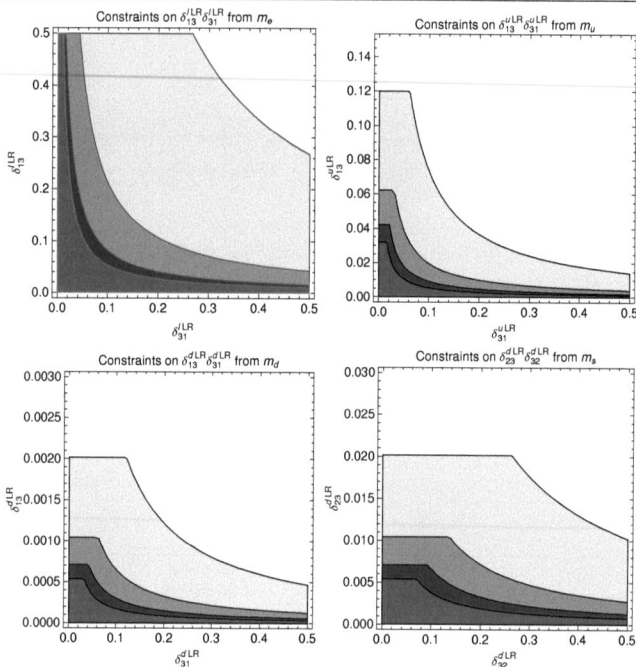

Figure 4.11: Upper left: Results of the two-loop contribution to the electron mass. Regions compatible with the naturalness principle for $M_{\text{SUSY}} = 200$ GeV (yellow), 500 GeV (green), 800 GeV (blue), 1000 GeV (red). Upper right to lower left: Results of the two-loop contribution to the up, down and strange quark mass. Region compatible with the naturalness principle (100% bound) for $M_{\text{SUSY}} = 500$ GeV (yellow), 1000GeV (green), 1500 GeV (blue), 2000 GeV (red).

5. CHIRALLY ENHANCED CORRECTIONS TO FCNC PROCESSES

In this chapter we study the effects of chirally-enhanced flavor-changing self-energies on FCNC in the generic MSSM. As discussed in chapter 3 there are two possibilities for a chiral enhancement. The first is a factor of $A_{ij}^d v_d/(M_{\text{SUSY}} \text{Max}[m_{d_i}, m_{d_j}])$ with a flavor-changing trilinear SUSY-breaking term A_{ij}^d dominating over a small quark mass $m_{d_{i,j}}$ in the denominator. M_{SUSY} is the mass scale determining the size of the loop diagram, i.e. M_{SUSY} is roughly the maximum of the gluino and squark masses running in the loop. The second possible chiral enhancement factor is $(v/M_{\text{SUSY}}) \tan \beta$ accompanied by a flavor-changing squark mass term involving squark fields of the same chirality. We have analyzed these self-energies in the context of charged-current processes in chapter 4 and this chapter contains the complementary study of FCNC processes. As an example, consider $B_s - \overline{B}_s$ mixing: In the presence of chirally enhanced corrections one must also take into account two- or even tree-loop diagrams, because the loop suppression is offset by the chiral enhancement factor (see Fig. 5.1). Similar corrections have been considered before in Ref. [19–22]. However, the authors of these papers have used a different definitions of the super-CKM basis and of the parameters $\delta_{fi}^{q\,AB}$ describing the flavour violation in the squark mass matrices. As a consequence, our results are hardly comparable to the ones obtained in Ref. [19–22]. We elaborate on the differences between Ref. [19–22] and our analysis in the next section. We note that models in which CKM elements and light-fermion masses are generated radiatively [66] (see chapter 6) require large trilinear SUSY-breaking terms. In the presence of these large A-terms (or of a large factor of $m_b \mu \tan \beta$ in combination with chirality-conserving flavor violation) it is important to include the effect of the chirality-flipping self-energies into FCNC processes. We will accomplish this in the next section by renormalizing the quark-squark-gluino vertex by a matrix-valued quark-field rotation in flavor space. In Sect. 5.2 the radiative decay $b \to s\gamma$ is examined in detail. The chirally enhanced corrections are only relevant for the large-$\tan \beta$ case here (or if $v_d A_{33}^d$ large). In Sect. 5.3, $\Delta F = 2$ processes are investigated, where large corrections occur irrespective of the size of $\tan \beta$

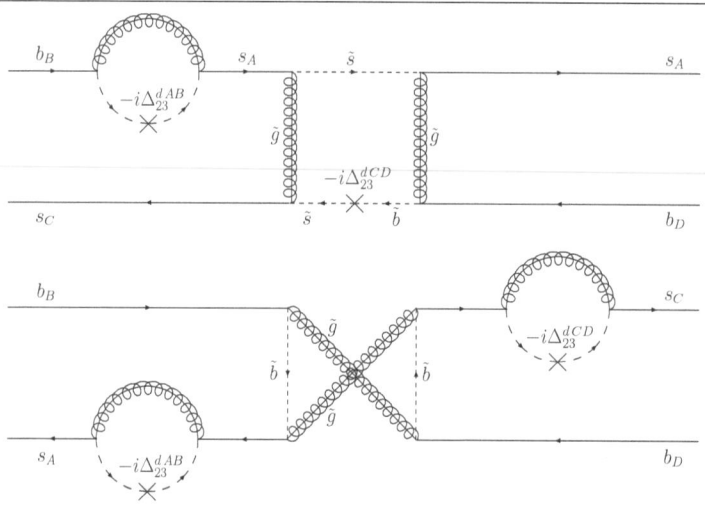

Figure 5.1: Examples of chirally enhanced two-loop and three-loop diagrams contributing to B_s mixing which can compete with (or even dominate over) the one-loop diagrams.

if the flavor violation is due to A-terms and if the squarks are not degenerate. Up-to-date measurements and theoretical Standard-Model predictions are used. The theoretical uncertainties of the SM are treated in a consistent and systematic way. In each case we compare the size of the FCNC transition to the previously known LO result: With the inclusion of our self-energies the bounds on the off-diagonal elements of the squark mass matrix change drastically, especially if the SUSY particles are rather heavy.

5.1. One-loop renormalization of the quark-squark-gluino vertex

In this section we show how to treat the chirally enhanced parts of flavor-changing self-energies in the full MSSM and how to absorb them into mixing matrices entering the Feynman rule of the quark-squark-gluino vertex. In this way all enhanced two-loop and three-loop diagrams are automatically included in the LO calculation to all orders in v/M_{SUSY}. Chirally enhanced corrections, in which we are interested here, have been calculated

5.1 One-loop renormalization of the quark-squark-gluino vertex

in Ref. [19–22] from a loop-corrected quark mass matrix, in a formalism in which the heavy SUSY particles are integrated out. When these expressions are combined with some one-loop amplitude this method can, in principle, reproduce the chirally enhanced two-loop (or three-loop) FCNC amplitude to leading non-vanishing order in $v/M_{\rm SUSY}$, where $v = \sqrt{v_u^2 + v_d^2} = 174\,\text{GeV}$ is the electroweak vev. Since the squark mixing angles scale as $v/M_{\rm SUSY}$, scenarios with large left-right mixing among squarks are not properly covered in this approach. In particular, the widely-studied large-$\tan\beta$ scenarios typically involve large sbottom mixing and therefore call for an analysis beyond the decoupling limit. Another important difference between Ref. [19–22] and this paper concerns the definitions of the super-CKM basis and the mass insertion parameters:

- We use the tree-level definition of the super-CKM basis which permits an analytical solution to the necessary all-order resummations of enhanced corrections. With the on-shell definition of Ref. [19–22] the self-energies and the squark mass matrices depend mutually on each other, which requires an iterative numerical evaluation of both quantities. This definition clouds the relations between observables and fundamental parameters like the trilinear SUSY-breaking terms $A_{fi}^{u,d}$ (as discussed later in this section). Our results are more transparent and enable us to identify a previously neglected parameter region with dramatically enhanced corrections to meson-antimeson mixing (see Sect. 5.3.)

- Concerning the definition of the mass insertion parameters $\delta_{fi}^{q\,AB}$ we choose the most common one (see for example [10–14, 16, 23–25, 75]) in which the mass insertion is the entire off-diagonal element of the squark mass matrix divided by the average squark mass. On the other hand, the authors of Ref. [19–22] only include the term $v_q A^q$ into their definition of the mass insertion parameters. This leads to an artificial dependence of all corrections, even the ones independent of a quark mass, on $\tan\beta$. Furthermore, if the authors of Ref. [19–22] would have chosen our definition of the mass insertions (while keeping their definition of the super-CKM basis) all chirally enhanced corrections would be simply absent, instead they would be implicitly contained in the definition of the $\delta_{fi}^{q\,AB}$'s.

As already mentioned, the chirally enhanced parts of the SQCD self-energies do not decouple. This means they do not vanish in the limit $M_{\rm SUSY} \to \infty$ but rather converge to a constant. Therefore the corresponding FCNC diagrams of the type in Fig. 5.1 scale with $M_{\rm SUSY}$ in the same way as the LO diagram. We will find that the two-loop and three-loop diagrams compete with the LO ones, especially for rather large and non-degenerate squark masses. There are different possibilities to handle flavor-changing self-energies in external legs: The conceptually simplest version is to treat them in the same way as one-particle-irreducible diagrams [29] as already done in chapter 3. So far we have implied this method, which is best suited to identify chirally enhanced contributions. Calculations are easiest, however, if one absorbs the chirally enhanced corrections into the quark-squark-gluino vertex through the unitary rotations ΔU_A^q of (3.20). These rotations alter the Feynman rule for the squark-quark-gluino vertex:

$$W_{s,i}^{q*} \to W_{s,j}^{q*} \left(1 + \Delta U_{Lji}^q\right)$$
$$W_{s,i+3}^{q*} \to W_{s,j+3}^{q*} \left(1 + \Delta U_{Rji}^q\right) \tag{5.1}$$

The procedure in (5.1) can be viewed as a short-cut to include the self-energy in the external quark line, in the spirit of Ref. [29]. Alternatively (5.1) can be interpreted as a finite matrix-valued renormalization of the quark fields which cancels the external self-energies and reappears in the Feynman rule of the quark-squark-gluino vertex.[1] The inclusion of the enhanced corrections into the LO calculation is now simply achieved by performing the replacements of (5.1) in this Feynman rule. Therefore, here the exact diagonalization of the squark mass matrix is preferred over the mass insertion approximation. The exact diagonalization has also the advantage that the analysis can be extended to the large $\tan\beta$ region in which certain off-diagonal entries can have the same size as the diagonal ones.
Here a comment on the definition of the super-CKM basis and the renormalization scheme is in order (see also section 3.4):

i) Recall that we defined the super-CKM basis as follows: Starting from some weak basis we diagonalize the tree-level Yukawa couplings and apply this unitary transformation to the whole supermultiplet. This is a natural definition of the super-CKM basis because of the direct correspondence between the SUSY-breaking Lagrangian and physical observables. Whenever we refer to some element of a squark mass matrix,

---

[1] We stress that we do not introduce ad-hoc counter-terms to the quark-squark-gluino vertex. Supersymmetry links the renormalization of the latter to the quark-quark-gluon vertex (which is unaffected by the rotations in (3.20)) and the renormalization of the soft SUSY-breaking terms (which can feed into the renormalization of the squark rotation matrices). The requirement to maintain the structure of a softly broken SUSY theory within the renormalization process restricts the allowed counterterms to the quark-squark-gluino vertex. Counterterms stemming from field renormalizations, however, are harmless in this respect, because field renormalizations trivially drop out from the LSZ formula for transition matrix elements.

this element is defined in this basis. When passing from LO to NLO or even higher orders the definition of the squark mass matrices is unchanged, i.e. no large chirally enhanced rotations appear at this step. Any additional non-enhanced (i.e. ordinary SQCD) corrections are understood to be renormalized in a way which amounts to a minimal renormalization of the squark mass matrices.

ii) From given squark mass matrices we calculate the self-energies $\Sigma_{fi}^{q\,RL}$ and then the rotations $\Delta U_{L,R}^q$ in (5.1). Calculating LO amplitudes with the corrected W_{sk}^{q*} from (5.1) then properly includes the desired chirally enhanced effects. The order of the two steps is important: First the super-CKM basis is defined from the tree-level structure of the Yukawa sector and the finite loop effects are included afterward, without influence on the definition of the super-CKM basis.

Alternatively one could define the super-CKM basis using an on-shell scheme which eliminates the self-energies in the external legs by shifting their effect into the definition of the super-CKM basis: Applying the inverse of the rotation in (5.1) first to the whole (s)quark superfields will leave the squark-quark-gluino vertex flavor-diagonal. (Further supersymmetry is still manifest, e.g. the sbottom field is the superpartner of the bottom field. This would not be the case if different rotations were applied to quark and squark fields.) If one defines this basis as the super-CKM basis (which now changes in every order of perturbation theory) one will find very different constraints on the off-diagonal elements of the squark mass matrices than with our method. The effect of the enhanced self-energies will be entirely absorbed into the values of the elements of the squark mass matrices, these self-energies will not appear explicitly, and the calculation of LO diagrams in the usual way will be sufficient. However, the squark mass matrices determined from data using this method will not be simply related to a mechanism of SUSY breaking, because the extracted numerical value of a given matrix element will also contain the physics associated with the chirally enhanced self-energies. Effects from SUSY breaking and electroweak breaking are interwoven now and further the elements of the squark mass matrix do not obey simple RG equations anymore.

It is illustrative to consider the popular case of soft-breaking terms which are universal at a high scale, say, the GUT scale. The unitary rotations diagonalizing the Yukawa couplings will lead to soft-breaking terms which are proportional to the unit matrix in flavor space. The RG evolution down to low scales will then lead to small flavor-off-diagonal LL entries of the squark mass matrices which are governed by the tree-level CKM matrix. Obviously, the elements of the squark mass matrices defined in this way are the quantities which one wants to probe in order to discriminate between high-scale universality and other possible mechanisms of soft flavor violation. The above-mentioned unitary rotations diagonalize the tree-level Yukawa couplings, while the rotations with $\Delta U_{L,R}^q$ are only meaningful after electroweak symmetry breaking and are therefore a low-scale phenomenon. During the RG evolution the Yukawa couplings essentially stay diagonal and the small unitary rotations bringing the low-scale Yukawa couplings back to diagonal form are unrelated to the soft breaking sector (and involve no chiral enhancement). Therefore the procedure described in items i) and ii) is the adequate method to probe the flavor structure of SUSY breaking. For a discussion of renormalization schemes in the context of MFV see Ref. [76].

Consider an FCNC transition $d_i \to d_f$ in MIA: If the squark mass $m_{\tilde{d}_i}$ is degenerate with $m_{\tilde{d}_f}$, the renormalization effects of the squark-quark-gluino vertex drop out in FCNC processes. This can be understood in the diagrammatical approach by realizing that diagrams with a flavor-changing self-energy in the outgoing f leg cancel with the diagram where the self-energy is in the incoming i leg, because the intermediate loop is the same for both diagrams. Thus, in order to demonstrate the effects of the renormalized quark-squark-gluino vertex non-degenerate squarks are necessary. For definiteness we choose the flavor-diagonal left-handed and right-handed mass terms equal and further set $m_{\tilde{q}1,\tilde{q}2} = 2m_{\tilde{q}3}$ if not mentioned otherwise. Lighter third-generation squarks are plausible in scenarios with high-scale flavor universality, in which renormalization group (RG) effects usually drive the bilinear soft terms of the third generation down.

At this point we may compare our results in (3.20) with the corresponding expressions in Ref. [19–22]. Unlike our $\Delta U_{L,R}^q$ the enhanced corrections in Ref. [19–22] depend explicitly on $\tan \beta$. This feature reflects the different definitions of the $\delta_{fi}^{q\,AB}$'s adopted in the two approaches. In particular, the constraints which we will derive from $\Delta F = 2$ processes in Sect. 5.3 are very different from those in Ref. [19–22].

5.2. $B \to X_s \gamma$

In this section we examine the radiative decay $b \to s\gamma$. We show how the renormalization of the quark-squark-gluino vertex affects the branching-ratio for different values of μ.[2] Throughout this section we assume that μ and the elements $\Delta_{ij}^{q\,AB}$ with $AB = LL, LR, RL, RR$, of the squark mass matrices are real. We consider only the case

---

[2]We find agreement of our LO Wilson coefficients with the gluino part given in Ref. [15, 66].

5.2 $B \to X_s \gamma$

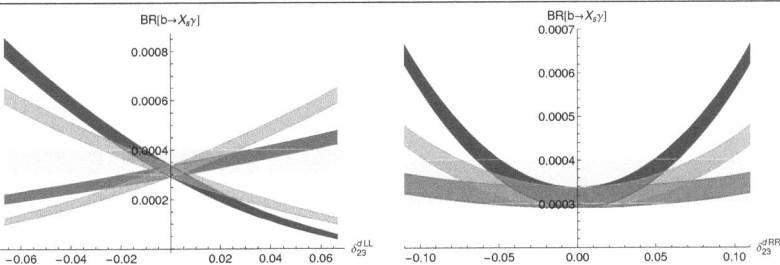

Figure 5.2: Br[$B \to X_s\gamma$] as a function of $\delta_{23}^{d\,LL}$ and $\delta_{23}^{d\,RR}$, respectively, for $m_{\tilde{g}} = 750\,\text{GeV}$, $m_{\tilde{q}1,2} = 2m_{\tilde{q}3} = 1000\,\text{GeV}$ and $\tan\beta = 50$. Yellow(lightest): experimentally allowed range for Br[$B \to X_s\gamma$]. Green, red, blue (light to dark): theoretically predicted range for $\mu/(1 + \Delta_b) = \pm 600\,\text{GeV}$ without renormalization, $+600\,\text{GeV}$ with renormalized vertices, $-600\,\text{GeV}$ with renormalized vertices. In the left figure the green band with positive (negative) slope corresponds to the LO branching-ratio with $\mu/(1 + \Delta_b) = +600\,\text{GeV}$ ($\mu/(1 + \Delta_b) = -600\,\text{GeV}$).

in which $|\mu| \tan\beta$ is large, because otherwise our new contributions are suppressed and no new constraints can be found. The reason for this feature is the chirality structure of the magnetic transition mediated by the operator O_7: Both the flavor-changing self-energy and the now flavor-conserving magnetic loop are necessarily chirality-flipping. The chiral enhancement of the latter is achieved by a large value of $|\mu| \tan\beta$. We further recall that the contributions from the dimension-six operators $O_{7b,\tilde{g}}$ and $O_{8b,\tilde{g}}$ (defined according to Ref. [66]) are suppressed by a factor of $\frac{M_{\text{SUSY}}}{\mu \tan\beta}$ compared to the contributions from $O_{7\tilde{g}}$ and $O_{8\tilde{g}}$. Furthermore, since all other SQCD contributions are also suppressed we only need to consider the magnetic operators and their chromomagnetic counterparts:

$$O_{7\tilde{g}} = eg_s^2(Q)\,\bar{s}\sigma_{\rho\nu}P_R b\, F^{\rho\nu}, \qquad O_{8\tilde{g}} = g_s^3(Q)\,\bar{s}\sigma_{\rho\nu}T^a P_R b\, G^{a\rho\nu} \qquad (5.2)$$

$$\tilde{O}_{7\tilde{g}} = eg_s^2(Q)\,\bar{s}\sigma_{\rho\nu}P_L b\, F^{\rho\nu}, \qquad \tilde{O}_{8\tilde{g}} = g_s^3(Q)\,\bar{s}\sigma_{\rho\nu}T^a P_L b\, G^{a\rho\nu} \qquad (5.3)$$

In [66] the matching scale Q is chosen as $Q = M_W$. We use $Q = m_t$ instead, because it is closer to the SUSY scale while still permitting 5-flavor running of α_s. The experimental value of [67] is taken at 2σ confidence level. For the theoretical prediction, the value of reference [77] is used at the lower and upper end of the error range. We have not used the cumbersome NNLO formula of Ref. [77], but have instead fitted C_{7SM} in the simple LO formula to reproduce the numerical NNLO result for $\Gamma(b \to s\gamma)$. The LO expression reads

$$\Gamma(b \to s\gamma) = \frac{m_b^5 G_F^2 |V_{tb}V_{ts}^*|^2 \alpha}{32\pi^4} \left(|C_7|^2 + \left|\tilde{C}_7\right|^2 \right)$$

$$C_7 = \frac{-16\sqrt{2}\pi^3 \alpha_s(\mu_b)}{G_F V_{tb}V_{ts}^* m_b} C_{7\tilde{g}} + C_{7SM} \tag{5.4}$$

$$\tilde{C}_7 = \frac{-16\sqrt{2}\pi^3 \alpha_s(\mu_b)}{G_F V_{tb}V_{ts}^* m_b} \tilde{C}_{7\tilde{g}}$$

To check our approximations we have also calculated the NLO evolution with matching at $Q = M_{\rm SUSY}$, but found only a slightly different result.

We now discuss the dependence of $b \to s\gamma$ on the different squark mass parameters $\Delta_{23}^{d\,AB}$ (or, equivalently, on the usual dimensionless quantities $\delta_{23}^{d\,AB}$): If the chirality-conserving elements $\Delta_{23}^{d\,LL,RR}$ are the non-minimal source of flavor violation $b \to s\gamma$ depends very strongly on $\mu\tan\beta$ already at the one-loop level (i.e. without the renormalization of the quark-squark-gluino vertex). With the inclusion of the flavor-changing self-energies in the external legs $C_{7\tilde{g},8\tilde{g}}$ is enhanced (suppressed) if μ is negative (positive) compared to the LO coefficient. The size of the effect is rather different for $\delta_{23}^{d\,LL}$ and $\delta_{23}^{d\,RR}$, because only in the first case interference with C_{7SM} is possible (see Fig. 5.2).

For the chirality-violating elements of the squark mass matrix, $\Delta_{23}^{d\,LR,RL}$, this dependence on $\mu\tan\beta$ is absent at LO and comes only into the game by the renormalization of the squark-quark-gluino vertex. Again the behavior is different for $\delta_{23}^{d\,LR}$ compared to $\delta_{23}^{d\,RL}$, since only in the first case interference with C_{7SM} is possible (see Fig. 5.3).

As we easily see from Fig. 5.3 the inclusion of the two-loop effects can substantially change the branching ratio. For positive (negative) values of μ the size of the Wilson coefficient $C_{7\tilde{g}}$ decreases (increases), i.e. the qualitative effect of our corrections is the same as for the LL and RR elements in Fig. 5.2.

In Fig. 5.4 we plot the constraints obtained from $\mathrm{Br}[B \to X_s\gamma]$ on $\delta_{23}^{d\,AB}$ versus $\mu/(1+\Delta_b)$ for different values of $m_{\tilde{g}}$. ($\Delta_b = \Delta_{d_3}$ is defined in (3.28).) All four different chirality combinations are shown. The constraints on $\delta_{23}^{d\,LR}$ and $\delta_{23}^{d\,LL}$ are stronger than the ones on $\delta_{23}^{d\,RL}$ and $\delta_{23}^{d\,RR}$ with exception of the "conspiracy" regions where the SUSY contributions overcompensate the SM value for C_7. For $\delta_{23}^{d\,LR,RL}$ the allowed region widens from bottom to top, meaning that negative values of μ strengthen the bounds on these quantities while

5.3 ΔF = 2 processes

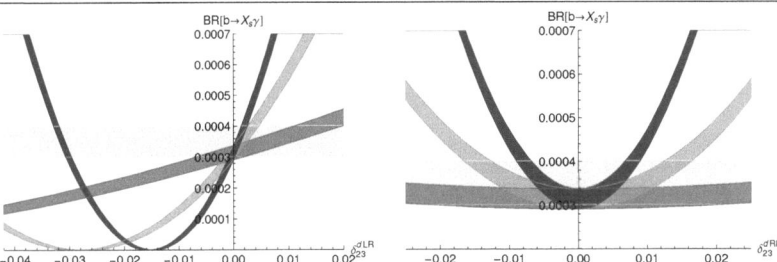

Figure 5.3: $\mathrm{Br}[\mathrm{B} \to \mathrm{X_s}\gamma]$ as a function of $\delta_{23}^{d\,LR}$ and $\delta_{23}^{d\,RL}$, respectively, for $m_{\tilde{g}} = 750\,\mathrm{GeV}$, $m_{\tilde{q}1,2} = 2m_{\tilde{q}3} = 1000\,\mathrm{GeV}$, and $\tan\beta = 50$. Yellow(lightest): experimentally allowed range for $\mathrm{Br}(\mathrm{B} \to \mathrm{X_s}\gamma)$. Green, red, blue (light to dark): theoretically predicted range for $\mu/(1+\Delta_b) = 0\,\mathrm{GeV}$, $\mu/(1+\Delta_b) = 600\,\mathrm{GeV}$, $\mu/(1+\Delta_b) = -600\,\mathrm{GeV}$.

positive values of μ weaken them. For $\delta_{23}^{d\,LL,RR}$ the effect of μ is different, as one can verify from the two lower plots in Fig. 5.4: The bounds on $\delta_{23}^{d\,LL,RR}$ always get stronger for increasing $|\mu|$ but are more stringent for $\mu < 0$ than for $\mu > 0$.

5.3. ΔF = 2 processes

In this section, we consider B_d, B_s, and K mixing. We show that the enhanced effects of the renormalized quark-squark-gluino vertices vanish for degenerate squark masses. However, if the squarks are non-degenerate our (N)NLO corrections are even dominant in a large region of the parameter space. In this analysis we consider complex $\delta_{ij}^{d\,AB}$'s to exploit the data on CP asymmetries. The effective Hamiltonian is given by

$$H_{\text{eff}}^{\text{SUSY}} =$$

$$\frac{-\alpha_s^2}{216}\Bigg[\ V_{s\,23}^{d\,LL}V_{t\,23}^{d\,LL}\left[24m_{\tilde{g}}^2 D_0\left(m_{\tilde{d}_s}^2,m_{\tilde{d}_t}^2,m_{\tilde{g}}^2,m_{\tilde{g}}^2\right)+66D_2\left(m_{\tilde{d}_s}^2,m_{\tilde{d}_t}^2,m_{\tilde{g}}^2,m_{\tilde{g}}^2\right)\right]Q_1$$

$$+V_{s\,23}^{d\,RR}V_{t\,23}^{d\,RR}\left[24m_{\tilde{g}}^2 D_0\left(m_{\tilde{d}_s}^2,m_{\tilde{d}_t}^2,m_{\tilde{g}}^2,m_{\tilde{g}}^2\right)+66D_2\left(m_{\tilde{d}_s}^2,m_{\tilde{d}_t}^2,m_{\tilde{g}}^2,m_{\tilde{g}}^2\right)\right]\tilde{Q}_1$$

$$+V_{s\,23}^{d\,LL}V_{t\,23}^{d\,RR}\left[504m_{\tilde{g}}^2 D_0\left(m_{\tilde{d}_s}^2,m_{\tilde{d}_t}^2,m_{\tilde{g}}^2,m_{\tilde{g}}^2\right)Q_4 - 72D_2\left(m_{\tilde{d}_s}^2,m_{\tilde{d}_t}^2,m_{\tilde{g}}^2,m_{\tilde{g}}^2\right)Q_4\right.$$

$$\left.+24m_{\tilde{g}}^2 D_0\left(m_{\tilde{d}_s}^2,m_{\tilde{d}_t}^2,m_{\tilde{g}}^2,m_{\tilde{g}}^2\right)Q_5 + 120D_2\left(m_{\tilde{d}_s}^2,m_{\tilde{d}_t}^2,m_{\tilde{g}}^2,m_{\tilde{g}}^2\right)Q_5\right]$$

$$+V_{s\,23}^{d\,LR}V_{t\,23}^{d\,LR}\left[204m_{\tilde{g}}^2 D_0\left(m_{\tilde{d}_s}^2,m_{\tilde{d}_t}^2,m_{\tilde{g}}^2,m_{\tilde{g}}^2\right)Q_2 - 36m_{\tilde{g}}^2 D_0\left(m_{\tilde{d}_s}^2,m_{\tilde{d}_t}^2,m_{\tilde{g}}^2,m_{\tilde{g}}^2\right)Q_3\right]$$

$$+V_{s\,23}^{d\,LR}V_{t\,23}^{d\,RL}\left[-132D_2\left(m_{\tilde{d}_s}^2,m_{\tilde{d}_t}^2,m_{\tilde{g}}^2,m_{\tilde{g}}^2\right)Q_4 - 180D_2\left(m_{\tilde{d}_s}^2,m_{\tilde{d}_t}^2,m_{\tilde{g}}^2,m_{\tilde{g}}^2\right)Q_5\right]\Bigg].$$

(5.5)

For definiteness we have quoted (5.5) for B_s mixing but the translation to other processes is trivial. The definitions of the operators can be found in Ref. [11, 14, 23]. We further used the following abbreviations:

$$V_{s\,fi}^{q\,RL} \equiv W_{f+3,s}^{\tilde{q}}W_{is}^{\tilde{q}*} \qquad V_{s\,fi}^{q\,LR} \equiv \sum_{j,k=1}^{3} W_{fs}^{\tilde{q}}W_{i+3,s}^{\tilde{q}*}$$

$$V_{s\,fi}^{q\,LL} \equiv \sum_{j,k=1}^{3} W_{fs}^{\tilde{q}}W_{is}^{\tilde{q}*} \qquad V_{s\,fi}^{q\,RR} \equiv \sum_{j,k=1}^{3} W_{f+3,s}^{\tilde{q}}W_{i+3,s}^{\tilde{q}*}$$

(5.6)

In the limit of two mass insertions and degenerate squark masses equation (5.5) simplifies to the result of [14] by substituting

$$D_{0,2}(m_{\tilde{q}_s},m_{\tilde{q}_t},m_{\tilde{g}},m_{\tilde{g}}) \to F_{0,2}(m_{\tilde{q}},m_{\tilde{q}},m_{\tilde{q}},m_{\tilde{q}},m_{\tilde{g}},m_{\tilde{g}})$$

$$V_{s\,fi}^{q\,LL} \to m_{\tilde{q}}^2\delta_{fi}^{d\,LL}, \quad V_{s\,fi}^{q\,RR} \to m_{\tilde{q}}^2\delta_{fi}^{q\,RR}, \quad V_{s\,fi}^{q\,RL} \to m_{\tilde{q}}^2\delta_{fi}^{q\,RL}, \quad V_{s\,fi}^{q\,LR} \to m_{\tilde{q}}^2\delta_{fi}^{q\,LR}.$$

(5.7)

(5.5) agrees with the result in Ref. [78] and corrects two color factors in Eq. II.9 of Ref. [9]. In Ref. [24], a NLO calculation of the effective $\Delta F = 2$ Hamiltonian has been carried out. The authors reduce the theoretical uncertainty and find corrections of about 15 percent to the LO result. They miss our effects from the flavor-changing self-energies, because they work with degenerate squark masses, so that the self-energy contributions cancel as discussed at the end of Sect. 5.1. As we will see, including the renormalized vertices can yield an effect of 1000% and more. So it is sufficient for our purpose to stick to the LO Hamiltonian of (5.5) with the renormalized vertices of Sect. 5.5. To incorporate the large logarithms from QCD we use the RG evolution computed in Refs. [14, 79]. For the bag factors parameterizing the hadronic matrix elements we take the lattice QCD values of Ref. [80]. We show the effects of the renormalization of the squark-quark-gluino vertex on $\Delta M_{d,s}$ and the general pattern of the new contributions in the following subsection.

A complete NLO calculation of supersymmetric QCD corrections to $\Delta F = 2$ Wilson coefficients with exact diagonalization of the squark mass matrices and non-degenerate squark masses was performed in [81]. However, our chirally enhanced effects are not included, because Ref. [81] adopts the definition of the super-CKM basis based on an on-shell definition of the quark fields, as described above. With this definition the chirally enhanced effects are implicitly contained in the numerical values of the $\delta_{ij}^{q\,AB}$. While the NLO corrections of Ref. [81] are typically numerically much smaller than ours, they are important to control the scale and scheme dependences of the LO diagrams. Therefore our results and those of Ref. [81] are complementary to each other.

5.3.1. $B-\bar{B}$ mixing

The amplitude of $B_q - \bar{B}_q$ mixing, $q = d$ or s, is conventionally denoted by M_{12}. New physics contributions will typically change magnitude and phase of this amplitude. $|M_{12}|$ is probed through the mass difference Δm_q among the two mass eigenstates of the $B-\bar{B}$ system, while any new contribution to $\arg M_{12}$ will modify certain CP asymmetries. For the formalism and phenomenology of $B-\bar{B}$ mixing we refer to Ref. [82], an update of the SM contributions to $B-\bar{B}$ mixing can be found in Ref. [83].

The chirally enhanced contributions are important for the constraints on $\delta_{ij}^{d\,LR} = \delta_{ji}^{d\,RL*}$. They are also relevant if one seeks constraints on $\delta_{ij}^{d\,LL}$ in the large-$\tan\beta$ region, but for this case $\mathrm{Br}[B \to X_s \gamma]$ is more powerful. Therefore we restrict our discussion in this section to the LR elements, for which our new effects lead to drastic changes in the SUSY-contributions to ΔM_q. We denote the result with renormalized squark-quark-gluino vertices by $\Delta M_{q,\mathrm{ren}}$. In Fig. 5.5 we show the ratio of $\Delta M_{q,\mathrm{ren}}$ to the LO result $\Delta M_{q,\mathrm{LO}}$, which is calculated from the gluino-squark box diagram without our new contributions. As one can easily see, the effects from the finite vertex renormalization drop out for degenerate squark masses, while dramatic effects for large and unequal squark masses are observed: For instance, a value of $\Delta M_{q,\mathrm{ren}}/\Delta M_{q,\mathrm{LO}} = 50$ implies that the constraint on the studied

$\delta_{ij}^{d\,AB}$ is stronger by a factor of $\sqrt{50} \approx 7$, because both $\Delta M_{q,\text{ren}}$ and $\Delta M_{q,\text{LO}}$ are practically quadratic in $\delta_{ij}^{d\,AB}$ (cf. MIA to see this). We remark that the value of $\tan \beta$ is inessential in this section, varying $\tan \beta$ leads to $\mathcal{O}(1\%)$ changes of our $\Delta F = 2$ results.

In order to determine the possible size of new physics (NP), it is necessary to know the SM contribution to the process in question. For $B - \overline{B}$ mixing, this first requires the control over hadronic uncertainties, which presently obscure the quantification of NP contributions from the precise data on ΔM_d [67] and ΔM_s [84, 85]. In the case of $B_d - \overline{B}_d$ mixing one must also address V_{td}, because $B_d - \overline{B}_d$ mixing is used to determine this CKM element through the usual fit to the unitarity triangle. A first analysis combining different quantities probing the $B_s - \overline{B}_s$ mixing amplitude has revealed a 2σ discrepancy of $\arg M_{12}$ with the SM prediction [83]. Since then the CKMfitter [86] and UTfit collaborations [87] have constrained the possible contributions of new physics to the $B_d - \overline{B}_d$ mixing and $B_s - \overline{B}_s$ mixing amplitudes with sophisticated statistical (Frequentist and Bayesian, respectively) methods, using new information on $\arg M_{12}$ gained from tagged $B_s \to J/\psi \phi$ data [88, 89]. We use the corresponding recent UTfit analysis as shown in Fig. 5.6.

The quantities C_{B_q} and ϕ_{B_q} shown in the plots are defined as:

$$C_{B_q} e^{2i\phi_{B_q}} = \Delta_q = \frac{\langle B_q | H_{\text{eff}} | \overline{B}_q \rangle}{\langle B_q | H_{\text{eff}}^{\text{SM}} | \overline{B}_q \rangle} = \frac{M_{12}}{M_{12}^{\text{SM}}} = \frac{|M_{12}^{\text{SM}}| + |M_{12}^{\text{NP}}| e^{2i\phi_{\text{NP}}}}{|M_{12}^{\text{SM}}|} \qquad (5.8)$$

Here ϕ_{NP} is the difference between the phase of the new physics contribution M_{12}^{NP} and the phase of the SM box diagrams. Refs. [83] and [91] show the experimental constraints in the complex Δ_q planes instead. The plots in Fig. 5.7 show the allowed regions in the complex $\delta_{23}^{d\,LR}$ and $\delta_{13}^{d\,LR}$ planes. The analogous constraints on the complex $\delta_{23}^{d\,RL}$ and $\delta_{13}^{d\,RL}$ planes look identical, because $|\Delta F| = 2$ processes are parity-invariant. To obtain Fig. 5.7 we have parameterized the border of the 95% CL region in Fig. 5.6 and determined the values $\text{Re}[\delta_{13,23}^{d\,AB}]$ and $\text{Im}[\delta_{13,23}^{d\,AB}]$ which correspond to this region by using (5.8) with M_{12}^{NP} calculated from $H_{\text{eff}}^{\text{SUSY}}$ in (5.5). The hadronic matrix elements are conventionally expressed in terms of the product of the squared decay constant $f_{B_q}^2$ and a bag factor. The dependence on f_{B_q} drops out in the ratio defining $C_{B_q} e^{2i\phi_{B_q}}$, which only involves the ratios of the bag factors of the different operators. That is, the sizable uncertainty of f_{B_q} does not enter at this step, but entirely resides in the allowed region for $C_{B_q} e^{2i\phi_{B_q}}$ plotted in Fig. 5.6. Therefore our results in Fig. 5.7 correspond to the ranges for $f_{B_q} \sqrt{B_q}$ (where B_q is the bag factor of the SM operator) used in (the web update of) Ref. [90]. These ranges are $f_{B_s} \sqrt{B_s} = (270 \pm 30)\,\text{MeV}$ and $f_{B_s} \sqrt{B_s}/(f_{B_d} \sqrt{B_d}) = 1.21 \pm 0.04$ (both at 1σ). In the case of $B_s - \overline{B}_s$ mixing the colored regions corresponding to different gluino masses do not overlap and do not contain the point $\delta_{23}^{d\,LR} = 0$, because the SM value M_{12}^{SM} is not in the 95% CL region of the UTfit analysis. The tension with the SM largely originates from the $B_s \to J/\psi\phi$ data [88, 89]. A large gluino mass suppresses the supersymmetric contribution to M_{12}, so that a larger value of Im $\delta_{23}^{d\,LR}$ is needed to bring $\arg M_{12}$ into the 95% CL region.

5.3.2. $K-\overline{K}$ mixing

In $K-\overline{K}$ mixing the situation is very different from $B-\overline{B}$ mixing, because the chiral enhancement factor A_{12}^d/m_s involves the small m_s rather than m_b. The observed smallness of FCNC transitions among the first two generations not only forbids large $\delta_{12}^{q\,AB}$ elements but also constrains the splittings among the squark masses of the first two generations severely. This observation suggests the presence of a $U(2)$ symmetry governing the flavor structure of the first two (s)quark generations. This symmetry cannot be exact, as it is at least broken by the difference $Y^s - Y^d$ of Yukawa couplings. That is, the numerical size of flavor-$U(2)$ breaking is somewhere between 10^{-4} and a few times 10^{-2}, depending on the size of $\tan\beta$. We may therefore fathom deviations from flavor universality in the same ballpark in the squark sector. Counting $\Sigma_{12}^{d\,RL}$ as first-order in some $U(2)$-breaking parameter, we realize that our chiral enhancement factors are of zeroth order in $U(2)$ breaking due to the appearance of the factor $1/m_s$ in (3.20). Therefore $K-\overline{K}$ mixing is extremely sensitive to the remaining source of flavor-$U(2)$ breaking in the problem, the mass splitting $m_{\tilde{q}_2} - m_{\tilde{q}_1}$. At this point we mention that it is important to control the renormalization of m_s in the presence of ordinary QCD corrections. In Appendix B of Ref. [47] it has been shown that all QCD corrections combine in such a way that the inverse power of m_s is the $\overline{\text{MS}}$ mass evaluated at the scale $Q = M_{\text{SUSY}}$, provided that gluonic QCD corrections to $\Sigma_{12}^{d\,RL}$ are also calculated in the $\overline{\text{MS}}$ scheme.

The sensitivity of the chirally enhanced corrections to the squark-mass splitting is displayed in Fig. 5.8. Constraints on $\delta_{12}^{d\,AB}$ from $K-\overline{K}$ mixing have been considered for a long time (see Refs. [10, 11]). Again we use the UTfit analysis (cf. the left plot of Fig. 5.9) exploiting the mass difference ΔM_K and the CP-violating quantity ϵ_K. We show our improved constraints on the complex $\delta_{12}^{d\,LR}$ element in the right plot of Fig. 5.9.

Figs. 5.8 and 5.9 are analogous to Figs. 5.5–5.7 addressing $B-\overline{B}$ mixing; we refer to the corresponding figure captions for further explanation. We find that $K-\overline{K}$ mixing indeed probes flavor violation in the squark sector of the first two generations at the per-mille level. The constraints on $\delta_{12}^{d\,LR}$ sharply grow with $|m_{\tilde{q}_2} - m_{\tilde{q}_1}|$.

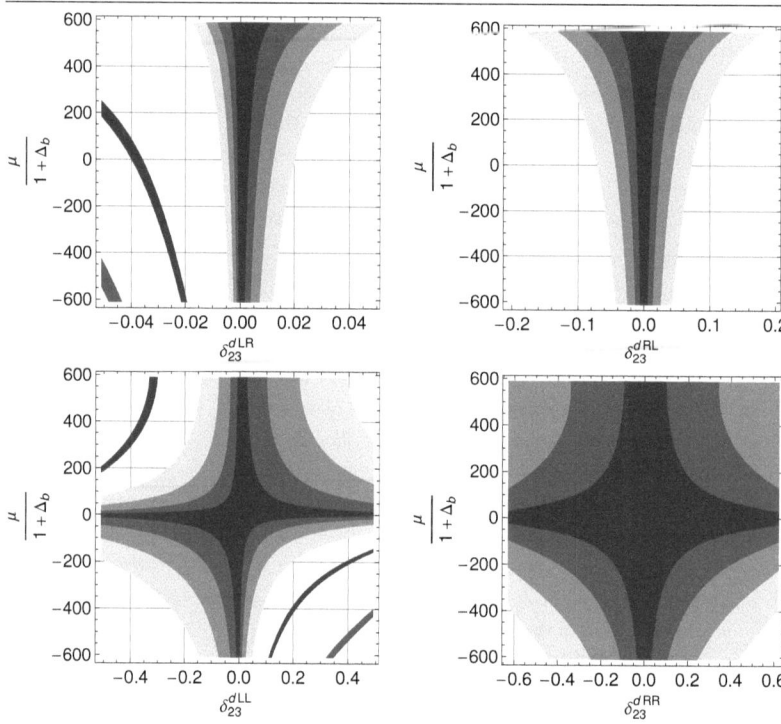

Figure 5.4: $B \to X_s\gamma$: allowed regions in the $\delta_{23}^{d\,LR,RL} - \frac{\mu}{1+\Delta_b}$ and $\delta_{23}^{d\,LL,RR} - \frac{\mu}{1+\Delta_b}$ planes for $m_{\tilde{q}1,2} = 2m_{\tilde{q}3} = 1000\,\text{GeV}$ and $\tan\beta = 50$. Yellow: $m_{\tilde{g}} = 2000\,\text{GeV}$, green: $m_{\tilde{g}} = 1500\,\text{GeV}$, red: $m_{\tilde{g}} = 1000\,\text{GeV}$, blue: $m_{\tilde{g}} = 500\,\text{GeV}$ (light to dark).

5.3 $\Delta F = 2$ processes

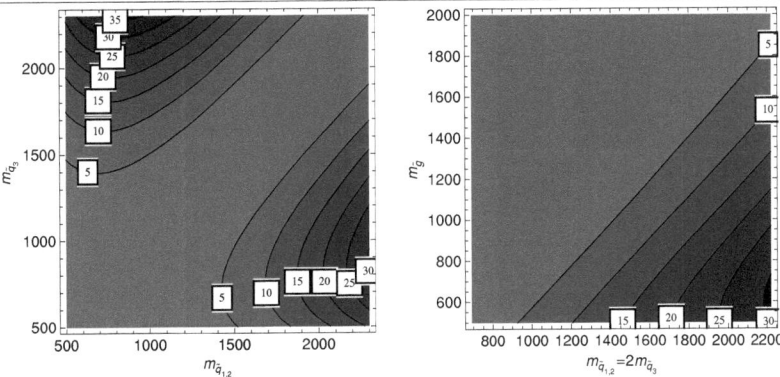

Figure 5.5: Left: Contour plot of $\Delta M_{q,\text{ren}}/\Delta M_{q,\text{LO}}$ as a function of $m_{\tilde{q}1,\tilde{q}2}$ and $m_{\tilde{q}3}$ with $m_{\tilde{g}} = 1000\,\text{GeV}$. Right: $\Delta M_{q,\text{ren}}/\Delta M_{BLO}$ as a function of $m_{\tilde{q}1,\tilde{q}2} = 2m_{\tilde{q}3}$ and $m_{\tilde{g}}$. The numbers in the little squares denote the value of $\Delta M_{q,\text{ren}}/\Delta M_{q,\text{LO}}$ of the corresponding contour.

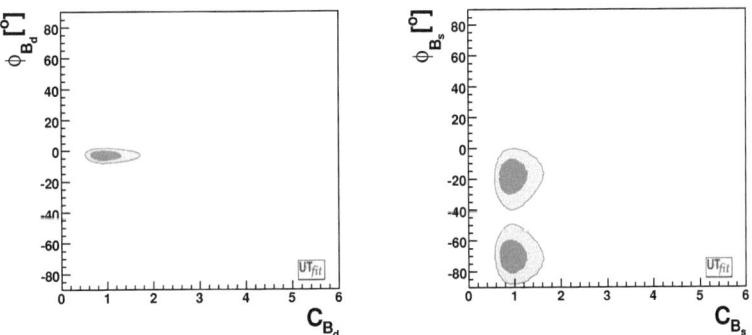

Figure 5.6: Allowed range for NP-contributions to B_q–$\overline{B}_q$ mixing, $q = d, s$, in the ϕ_{B_q}–C_{B_q} plane taken from the web update of Ref. [90]. See euqation (5.8) and related text for details. For a related CKMfitter analysis see Ref. [91].

5. Chirally enhanced corrections to FCNC processes

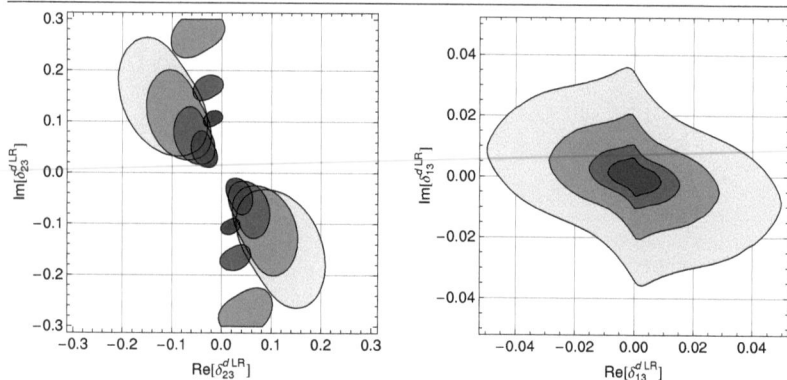

Figure 5.7: Allowed regions in the complex $\delta^{d\,LR}_{13,23}$-plane from B_s-mixing (left plot) and B_d-mixing (right plot) with $m_{\tilde{q}1,2} = 2m_{\tilde{q}3} = 1000\,\text{GeV}$. The yellow, green, red and blue (light to dark) areas correspond to the 95% CL regions for $m_{\tilde{g}} = 1200\,\text{GeV}$, $m_{\tilde{g}} = 1000\,\text{GeV}$, $m_{\tilde{g}} = 800\,\text{GeV}$, $m_{\tilde{g}} = 600\,\text{GeV}$. The effect is practically insensitive to $\tan\beta$. The same constraints are obtained for the complex $\delta^{d\,RL}_{13,23}$-plane.

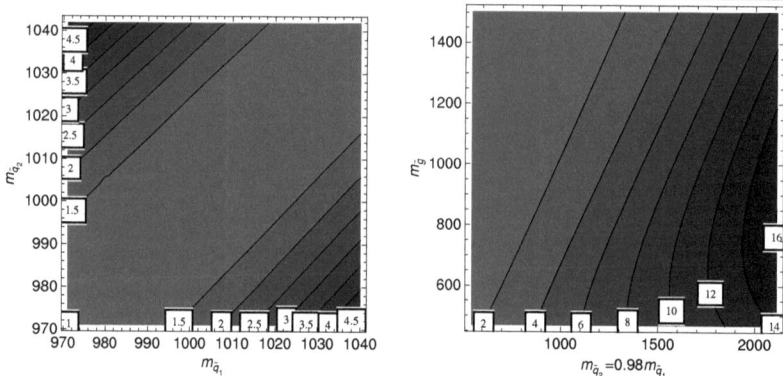

Figure 5.8: Left: Contour plot of $\Delta M_{K\,\text{ren}}/\Delta M_{K\,\text{LO}}$ as a function of $m_{\tilde{q}1}$ and $m_{\tilde{q}2}$ with $m_{\tilde{g}} = 1000\,\text{GeV}$. Right: $\Delta M_{K\,\text{ren}}/\Delta M_{K\,\text{LO}}$ as a function of $m_{\tilde{q}2} = 0.98 m_{\tilde{q}1}$ and $m_{\tilde{g}}$.

5.3 $\Delta F = 2$ processes

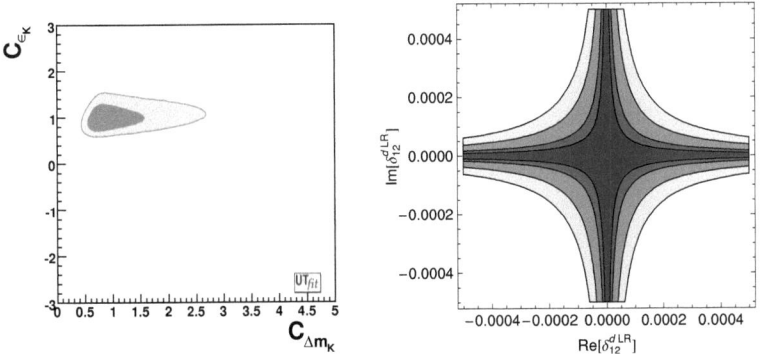

Figure 5.9: Left: Allowed range for NP contributions to K mixing determined by the UTFit collaboration [90]. The light region corresponds to 95% CL. Right: Allowed region in the complex $\delta_{12}^{d\,LR}$-plane with $m_{\tilde{g}} = 750\,\text{GeV}$, $m_{\tilde{q}1} = 1000\,\text{GeV}$. Yellow: $m_{\tilde{q}2} = 1000\,\text{GeV}$, green: $m_{\tilde{q}2} = 999\,\text{GeV}$, red: $m_{\tilde{q}2} = 998\,\text{GeV}$, blue: $m_{\tilde{q}2} = 997\,\text{GeV}$ (light to dark). The same constraints are found for $\delta_{12}^{d\,RL}$.

6. Radiative generation of light fermion masses

In chapter 3 we have seen that chirally-enhanced self-energies with sfermions and gauginos as virtual particles significantly correct the relations between the physical masses and the Yukawa couplings. Furthermore, we have derived very strong bounds on the mass-insertions $\delta^{f\,LR}_{11,22}$ by requiring that the supersymmetric corrections do not exceed the measured masses. This suggests the idea that the light fermion Yukawa couplings might be zero at tree-level and are generated radiatively via loops involving the trilinear A terms. In this chapter we want to study the phenomenological implications of such a model with radiative mass generation taking into account the chirally enhanced corrections to FCNC processes as described in section 3.4 and chapter 5.

6.1. Description of the model

The smallness of the Yukawa couplings of the first two generations suggests that these couplings are generated through radiative corrections [66, 92–96]. In the context of supersymmetric theories these loop-induced couplings arise from diagrams involving squarks (sleptons) and gluinos (binos) for quarks (leptons)[1] . Of course, the heaviness of the top quark requires a special treatment of Y^t and the successful bottom-tau Yukawa unification suggests to keep the tree-level Yukawa couplings for the third generation. At large $\tan\beta$, this idea gets even more support from the successful unification of the top and bottom Yukawa coupling, as suggested by some SO(10) GUTs. In the modern language of Refs. [97, 98] the global $[U(3)]^5$ flavor symmetry of the gauge sector (here we do not consider neutrinos) is broken down to $[U(2)]^5 \times [U(1)]^2$ by the Yukawa couplings of the third generation. Here the five $U(2)$ factors correspond to rotations of the left-handed doublets and the right-handed singlets of the first two generation quarks and leptons in flavor space,

---
[1]Of course also the bino diagram conributes to the quark masses, but it is supressed by a factor $\frac{3\alpha_1}{8\alpha_s}$

6.1 Description of the model

respectively. This means we have

$$Y^f = \begin{pmatrix} 0 & 0 & 0 \\ 0 & 0 & 0 \\ 0 & 0 & y^f \end{pmatrix}, \quad V^{(0)} = \begin{pmatrix} 1 & 0 & 0 \\ 0 & 1 & 0 \\ 0 & 0 & 1 \end{pmatrix} \quad (6.1)$$

in the tree-level Lagrangian[2]. We next assume that the soft breaking terms $\Delta_{ij}^{\tilde{q}\,LL}$ and $\Delta_{ij}^{\tilde{q}\,RR}$ possess the same flavor symmetry as the Yukawa sector, which implies that $\mathbf{M}_{\tilde{q}}$, $\mathbf{M}_{\tilde{d}}$ and $\mathbf{M}_{\tilde{u}}$ are diagonal matrices with the first two entries being equal. For transitions involving the third generation the situation is different because flavor violation can occur not only because of a misalignment between A^u and A^d but also due to a misalignment with the Yukawa matrix. So the elements $A_{j3}^{u,d}$ do not only generate the CKM matrix at one-loop, they also act as a source of non-minimal flavor violation and thus are constrained by FCNC processes.

This model has several advantages compared to the generic MSSM:

- Flavor universality holds for the first two generations. Thus our Model is minimally flavor-violating according to the definition of [97] with respect to the first two generations since the quark and the squark mass matrices are diagonal in the same basis. This provides an explanation of the precise agreement between theory and experiment in K and D physics. However, double mass insertions affect transitions between the first two generations (see next section).

- The SUSY flavor problem is reduced to the quantities $\delta_{13,23}^{q\,RL}$. However, these flavor-changing elements are less constrained from FCNCs and might even explain a possible new CP phase indicated by recent data on B_s mixing. Furthermore, as we will see in chapter 8 these elements can also induce a right-handed W coupling which can explain discrepancies between inclusive and exclusive determinations of V_{ub} and V_{cb}.

- The flavor symmetry of the Yukawa sector protects the quarks of the first two generations from a tree-level mass term.

- The model is economical: Flavor violation and SUSY breaking have the same origin.

- Small quark masses and small off-diagonal CKM elements are explained by a loop suppression.

---
[2]We assume that the large mixing angles of the PMNS matrix do not stem from the charged leptons but rather from the neutrinos.

- The SUSY CP problem is substantially alleviated by an automatic phase alignment [66]. In addition, the phase of μ does not enter the EDMs at the one-loop level because the Yukawa couplings of the first two generations are zero.

6.2. Phenomenological consequences in the quark sector

Although the B factories have confirmed the CKM mechanism of flavor violation with very high precision, leaving little room for new sources of FCNCs, we want to show in this section that the possibility of radiative generation of quark masses and of the CKM matrix still remains valid. Even though our model is minimally flavor-violating regarding only the first two generations, this is no longer true if the third generation is involved. The reason for this is that the A-terms cannot be constructed out of the tree-level Yukawa coupling and vice versa as demanded by the definition of MFV given in Ref. [99]. While V_{us} and V_{cd} are solely generated by a misalignment between A^u and A^d the situation concerning the CKM elements containing the third generation is more involved since the top and bottom Yukawa couplings fix the rotations involving the third generation. Therefore, V_{ub}, V_{cb}, V_{td} and V_{ts} are generated by the misalignment between the loop-corrected Yukawa couplings Y^u_{eff} and Y^d_{eff}. In the following we will concentrate on the two simple limiting cases in which either A^d is diagonal (in the same basis as Y^d) and the CKM elements are generated by the off-diagonal elements of A^u or on the opposite case in which A^u is diagonal but A^d is not. For obvious reasons we will call theses scenarios "CKM generation in the down-sector" and "CKM generation in the up-sector", respectively. Even though the elements $\delta^{q\,RL}_{13,23}$ are not needed for the generation of the CKM matrix, they can in principle be different from zero. However, we will concentrate on the minimal case in which $\delta^{q\,LR}_{13,23}$ are the only sources of non-minimal flavor violation. Note that it is in principle also possible to generate the fermion masses with the non-holomorphic terms given in equation (2.8). Such a scenario (as proposed in Ref. [66, 100]) would then lead to additional effects in the Higgs sector.

6.2.1. CKM generation in the down-sector

If the CKM matrix is generated in the down sector, the off-diagonal elements $\Delta^{d\,LR}_{13,23}$ are determined by the requirement that they generate the observed CKM matrix. From equation (3.20) and (3.23) the off-diagonal elements are determined by:

$$\Sigma^{d\,LR}_{13} = m_b V_{ub} \tag{6.2}$$

$$\Sigma^{d\,LR}_{23} = m_b V_{cb} \tag{6.3}$$

6.2 Phenomenological consequences in the quark sector

Here the bottom quark mass has to be evaluated at the same scale as the self-energy. Since the off-diagonal elements needed to generate the CKM matrix in the down-sector are very small (see figure 4.5 and 4.4) the mass-insertion approximation excellently reproduces the exact result. Therefore, we can solve analytically for $\Delta_{13,23}^{d\,LR}$ by using equation (3.16).

In leading order of the MIA the flavor off-diagonal elements $\delta_{13,23}^{d\,LR}$ enter FCNC processes involving the third generation. Furthermore, also Kaon and D mixing are affected by diagrams containing the combination $\delta_{13}^{d\,LR} \times \delta_{32}^{d\,RL}$. Even though Kaon mixing is very sensitive to NP, especially to new sources of CP violation, the product $\delta_{13}^{d\,LR} \times \delta_{32}^{d\,RL}$ is too small to give sizable effects. The contribution to D mixing is even further suppressed since it is generated by a chargino diagram. However, $b \to s(d)\gamma$ is very sensitive to $\delta_{23}^{d\,LR}$ ($\delta_{13}^{d\,LR}$) since it is both flavor and chirality violating. Even though the relative effect (compared to the SM contribution) in $b \to s\gamma$ and $b \to d\gamma$ is approximately equal, $b \to s\gamma$ turns out to be the process which is most sensitive to radiative flavor violation stemming from the down sector since it is measured much more precisely than $b \to d\gamma$. The new contributions affect the Wilson coefficients C_7 and C_8 and the interference with the SM contribution is constructive. However, one has to take into account the chirally enhanced corrections discussed in the previous chapter. Therefore, also the gluino constraints depend on μ and $\tan\beta$ (see figure 6.1).

In principle, no symmetry argument forbids non-vanishing terms $A_{31,32,33}^d$ in our model. However, $\delta_{13,23}^{d\,RL}$ contribute to C_7' and C_8' and $\delta_{33}^{d\,RL}$ is already non-zero since it also contains $m_b\mu\tan\beta$. As we see from equation (5.4) the primed Wilson Coefficients do not interfere with the SM ones. Therefore, in the absence of sizable charged Higgs contributions, their effect on the branching ratio is suppressed. Furthermore, since the sign of $\delta_{13,23}^{d\,RL}$ is not fixed a destructive interference with the charged Higgs contribution is possible which could weaken the bounds on the charged Higgs mass (in addition to a possible chargino contribution).

6.2.2. CKM generation in the up-sector

If the CKM matrix is generated in the up-sector the off-diagonal elements are determined by:

$$\Sigma_{13}^{u\,LR} = -m_t V_{ub} \tag{6.4}$$

$$\Sigma_{23}^{u\,LR} = -m_t V_{cb} \tag{6.5}$$

Due to the large top mass these off-diagonal elements are much larger than in the case of CKM generation in the down sector. Therefore, already the requirement that the lighter

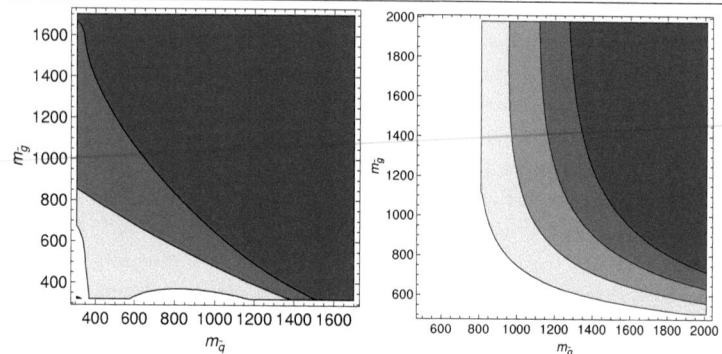

Figure 6.1: Allowed regions in the $m_{\tilde{g}} - m_{\tilde{q}}$ plane.
Left: Constraints from $b \to s\gamma$ for different values of $m_b \mu \tan\beta/(1+\Delta_b)$ assuming that the CKM matrix is generated in the down sector. Yellow(lightest): $m_b \mu \tan\beta/(1+\Delta_b) = 120\,\text{TeV}^2$, red: $m_b \mu \tan\beta/(1+\Delta_b) = 0\,\text{TeV}^2$ and blue: $m_b \mu \tan\beta/(1+\Delta_b) = -120\,\text{TeV}^2$.
Right: Constraints from Kaon mixing for different values of M_2 assuming that the CKM matrix is generated in the up sector. Yellow(lightest): $M_2 = 800\,\text{GeV}$, green: $M_2 = 600\,\text{GeV}$, red: $M_2 = 400\,\text{GeV}$ and blue: $M_2 = 200\,\text{GeV}$.

stop mass does not violate the bounds for direct searches requires the diagonal elements of the squark mass matrix to be heavier than approximately $(700\,\text{GeV})^2$. Furthermore, the mass-insertion approximation does not necessarily hold for such large off-diagonal elements. Therefore, one cannot solve analytically for $\Delta^{u\,LR}_{13,23}$ but rather has to determine these elements numerically. However, for squark masses above 700GeV the off-diagonal elements determined by the mass-insertion approximation turn out to be only up to ten percent bigger than the numerically obtained elements. Therefore, it is still possible to rely on the MIA for an qualitative understanding of the flavor structure.

If the CKM matrix is generated in the up-sector one would naively expect chargino contributions to $b \to s\gamma$, $b \to d\gamma$ and $B_{s,d}$ mixing. However, $B_{s,d}$ mixing does not give useful constraints on $\delta^{u\,LR}_{13,23}$ and $b \to s\gamma$ also heavily depends on μ and $\tan(\beta)$. Furthermore, we again have to take into account the chirally enhanced effects by using the effective chargino vertex of equation (3.34). In the present case of radiative flavor violation in the up-sector

6.2 Phenomenological consequences in the quark sector

these expression simplifies to:

$$\Gamma^{\tilde{\chi}_k^{\pm} L}_{d_i \tilde{u}_s} = \sum_{j=1}^{3} \begin{pmatrix} V_{ud} & V_{us} & 0 \\ V_{cd} & V_{cs} & 0 \\ 0 & 0 & 1 \end{pmatrix}_{ji} \left(V^{\tilde{\chi}^{\pm}*}_{k2} Y^{u_3} \delta_{j3} W^{\tilde{u}*}_{j+3,s} - g_2 V^{\tilde{\chi}^{\pm}*}_{k1} W^{\tilde{u}*}_{js} \right),$$

$$\Gamma^{\tilde{\chi}_k^{\pm} R}_{d_i \tilde{u}_s} = U^{\tilde{\chi}^{\pm}}_{k2} Y^{d_3} \delta_{i3} \sum_{j=1}^{3} \begin{pmatrix} V_{ud} & V_{us} & 0 \\ V_{cd} & V_{cs} & 0 \\ 0 & 0 & 1 \end{pmatrix}_{ji} W^{\tilde{u}*}_{js}. \tag{6.6}$$

Note that the Yukawa couplings of the light fermions are zero and also the CKM matrix elements connecting the third with the first two generations vanish for external down-type quarks. This is easy to understand since the tree-level Yukawa couplings of the light quarks are zero and the CKM matrix is induced in the up-sector meaning that the down quarks are not rotated by self-energies.

However, there are other, not so obvious, contribution to FCNC processes: Kaon and D mixing. An effective element $\delta^{u\,LL}_{12\,\text{eff}}$ is induced through the double mass insertion $\delta^{u\,LR}_{13} \times \delta^{u\,LR*}_{23}$. Note that this element is proportional to two powers of an electroweak vev and is therefore not subjected to the $SU(2)$ relation which connects the left-handed squarks. Therefore, on the one hand only chargino diagrams contribute to K mixing while on the other hand no chargino diagrams contributes to D mixing. However, since Kaon mixing is more sensitive to CP violation, and $\delta^{d\,LL}_{12\,\text{eff}}$ carries the CKM phase γ, these constraints turn out to be stronger than the constraints from D mixing. The allowed regions in the $m_{\tilde{q}}$-$m_{\tilde{g}}$ plane for different values of M_2 are shown in figure 6.1 b). Note that the constraints are nearly independent of μ and $\tan\beta$ since the quark-squark coupling is to the gaugino component of the charginos.

Another process which is sensitive to the combination $\delta^{u\,LR}_{13} \times \delta^{u\,LR*}_{23}$ via chargino loops is $K \to \pi\nu\nu$ [18, 101–103]. Even though, at present, this process does not give useful bounds, upcoming NA46 results will change this situation in the future. Figure 6.2 and 6.3 shows the predicted branching ratios for $K_L \to \pi\nu\bar{\nu}$ and $K^+ \to \pi^+\nu\bar{\nu}$. Note that the branching ratios are again to a very good approximation independent of μ and $\tan\beta$.

In principle $\delta^{u\,LR}_{23}$ also contributes to $B \to K\nu\bar{\nu}$ via a Z penguin (and at the same time also to $B_s \to \mu\mu$ which is strongly correlated to $B \to K\nu\bar{\nu}$ in the MSSM) [8, 26, 104]. Even

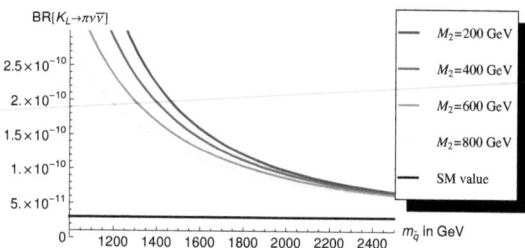

Figure 6.2: Predicted branching ratio for the rare Kaon decay $K_L \to \pi \nu \bar{\nu}$ assuming that the CKM matrix is generated in the up-sector for $m_{\tilde{q}} = m_{\tilde{g}}$. The branching ratio is enhanced due to a constructive interference with the SM contribution.

though the branching ratios are slightly enhanced, they also depend on A^t, μ and $\tan(\beta)$. Furthermore, $B \to K \nu \bar{\nu}$ is also correlated to $b \to s\gamma$ which forbids large effects [104].

Of course also in the up-sector no symmetry argument forbids non-zero elements $\delta^{u\,LR}_{31,32,33}$. While $\delta^{u\,LR}_{33}$ affects as already discussed $b \to s\gamma$ and $B \to K \nu \bar{\nu}$ the elements $\delta^{u\,LR}_{31,32}$ are rather separated from FCNC processes since they enter these processes only in combination with small quark masses and small chargino mixing. Their mere effect is to correct the sqaurk eigenvalues via diagonalization. However, as we will see in chapter 8, they can induce a sizeable right-handed W coupling if at the same time also $\delta^{d\,LR}_{33}$ is large.

6.3. Phenomenological consequences in the lepton sector

As already noted in section 6.1 our model solves the SUSY CP problem because the A-terms are real in the same basis as the Yukawa coupling of the light fermions. Furthermore, the phase of μ practically only enters at the two-loop level. Also the decay $\mu \to e\gamma$ receives no contributions in addition to the MFV ones. However, the anomalous magnetic moment of the muon (and to less extent of the electron) is affected due to the chirality violation in the slepton mass matrix:

The great triumph of the Dirac equation was the successful prediction of the magnetic

6.3 Phenomenological consequences in the lepton sector

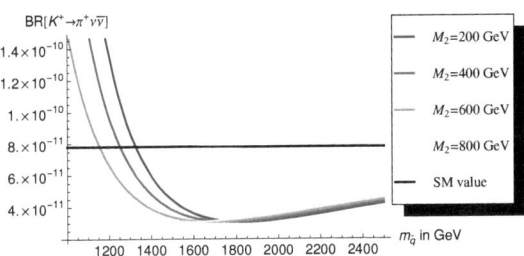

Figure 6.3: Predicted branching ratio for the rare Kaon decay $K^+ \to \pi^+ \nu \bar{\nu}$ assuming that the CKM matrix is generated in the up-sector for $m_{\tilde{q}} = m_{\tilde{g}}$. The branching ratio is enhanced for light SUSY masses but suppressed if the scale of SUSY breaking is higher.

moments of fermions:

$$\vec{\mu} = g_f \frac{e_f}{2m_f} \vec{S} \tag{6.7}$$

where $g_f = 2$ is the gyromagnetic ratio. However, loop corrections lead to deviations of the gyromagnetic factor from two. The magnetic dipole moment interaction relevant for this corrections is given as

$$\frac{ie}{2m_\mu} F(q^2) \bar{u}(p_f) \sigma_{\mu\nu} q^\mu \epsilon^\nu u(p_i) \tag{6.8}$$

where $q = p_f - p_i$ is the momentum and ϵ is the polarization vector of the external photon. The anomalous magnetic dipole moment of the muon is then given as $(g-2)_\mu = 2F(q^2 = 0)$. The deviation from two is defined as $a_\mu = \frac{(g-2)_\mu}{2}$. Hence, comparison of experiment and theory tests the SM (and its extensions) at its quantum loop level. In this contex, the anomalous magnetic moment of the muon is of special interest, because the discrepancy between the SM prediction and experiment is 3.1σ [105]:

$$a_\mu^{\text{exp}} - a_\mu^{\text{SM}} = (24.6 \pm 8.0) \times 10^{-10}. \tag{6.9}$$

In exact supersymmetric theories the gyromagnetic ratio for all fermions is exactly 2 [106]. Therefore, the anomalous magnetic moment of the muon directly probes SUSY breaking. A pleasant feature of supersymmetry, which singles it out with respect to other models, is that it would naturally lead to the observed deviation from the SM value [107–114]. The usual approach is to choose a suitable (large) value of the term $m_\mu \mu \tan\beta$ in the slepton mass matrix. In order to achieve the right value for the anomalous magnetic moment, the higgsino mass parameter μ must be positive and large values for $\tan\beta$, the ratio of the two vacuum expectation values, are favored. Even though large $\tan\beta$ scenarios are also motivated by the GUT relation $y_t = y_b$, problems in processes like $b \to s\gamma$, $B_{d,s} \to \mu\mu$ and $B \to (D)\tau\nu$ can occur, due to the parametric enhancement by $\tan\beta$. In supergravity, $B_{d,s} \to \mu\mu$ and the anomalous magnetic moment of the muon are correlated, limiting the possible size of a_μ^{SUSY} [115]. Therefore, if $\tan\beta$ is large, the Higgs has to be heavy in the constrained MSSM [116].

However, there exists also a second, less studied way in the MSSM how to account for the anomalous magnetic moment of the muon: The entry in the slepton mass matrix involving the trilinear A-term[3], A_{22}^l, can also reproduce the desired effect without influencing quark decays or the Higgs potential. This possibility is realized in our model with radiative generation of fermion masses [66]. Since in our model the trilinear A-terms are chosen in such a way that they generate the light fermion masses of the first and second generation, the contributon to the anomalous magnetic moment depends only on the slepton and the bino mass and is positive definite. In presence of flavor violating elements in the slepton mass matrix it is possible to generate the muon and electron mass radiatively via couplings involving y_τ. However the very same diagram where an additional photon is attached also contributes to the anomalous magnetic moment. No chargino diagram contributes due to the absence of a tree-level Yukawa coupling. Furthermore, we can neglect neutralino mixing between the bino and the neutral wino. Then the magnetic moment is given by:

$$a_\mu = m_\mu \frac{\alpha_1}{2\pi} M_1 \sum_{s=1}^{6} \Re\left(W_{2s}^{\tilde{\ell}*} W_{5s}^{\tilde{\ell}*}\right) m_{\tilde{\ell}_s}^2 D_0(M_1^2, m_{\tilde{\ell}_s}^2, m_{\tilde{\ell}_s}^2, m_{\tilde{\ell}_s}^2), \qquad (6.10)$$

where the rotation matrix $W^{\tilde{\ell}}$ and the masses $m_{\tilde{\ell}}$ and M_1 must fulfill the following condition:

$$m_\mu \stackrel{!}{=} \frac{\alpha_1}{4\pi} M_1 \sum_{s=1}^{6} W_{2s}^{\tilde{\ell}*} W_{5s}^{\tilde{\ell}*} B_0(M_1^2, m_{\tilde{\ell}_s}^2). \qquad (6.11)$$

For given diagonal elements of the slepton mass matrix, Eq. (6.11) is an implicit equation for the off-diagonal elements. The same rotation matrix $W^{\tilde{\ell}}$ must then be inserted in

---
[3]Vacuum stability requires either $\tan\beta \approx 1$ or the use of the non-analytic A' terms [66].

6.3 Phenomenological consequences in the lepton sector   71

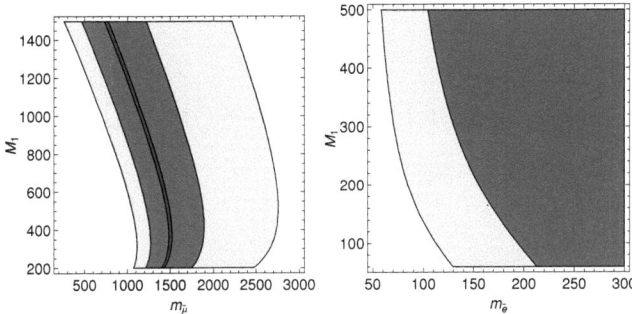

Figure 6.4: Left: Allowed region in the M_1-$m_{\tilde{\mu}}$ plane assuming that the muon Yukawa coupling is generated radiatively by $v_d A_{22}^l$. Here $m_{\tilde{\mu}}$ is the lighter smuon mass. Yellow(lightest): $a_\mu \pm 2\sigma$, red: $a_\mu \pm 1\sigma$, blue(darkest): a_μ.
Right: Allowed region in the M_1-$m_{\tilde{e}}$ plane assuming that the electron Yukawa coupling is generated radiatively by $v_d A_{11}^l$. Yellow(lightest): $a_\mu \pm 2\sigma$, red: $a_\mu \pm 1\sigma$.

Eq. (6.10). Assuming that the discrepancy in (6.9) can be explained with supersymmetry we can determine the allowed region in parameter space. The result is shown in the left plot of Fig. 6.4 where we see that a model with radiative generation of the muon mass predicts a smuon with masses approximately between 1 TeV and 4 GeV. This is a very clean and strong prediction which gains special importance in the light of the forthcoming LHC results: Since the LHC is only sensitive to light slepton $m_{\tilde{l}} \leq 400\,\text{GeV}$ a detection of a smuon would directly disprove the model with radiatively generated fermion masses. Stated positively, a light smuon would prove a non-zero tree-level Yukawa coupling in the MSSM and provide us with a lower bound. The same discussion applies as well to the electron and its Yukawa coupling. However, even though the anomalous magnetic moment of the electron is measured very precisely [117], it is used to determine α. Therefore, in order to use the anomalous magnetic moment of the muon in order to put bounds on NP we need an independent determination of α [72]. The second best way to measure the fine structure constant is from a Rubidium atom experiment [118]. Using these informations we can qualitatively make the same statements as in the muon case. However, quantitatively the constraints are weaker due to the smallness of the electron mass and the uncertainty coming from the second best measurement of α (see figure 6.4 b)).

7. LL MASS INSERTIONS AND CONSTRAINTS FROM K AND D MIXING

Already in the early stages of minimal supersymmetric standard model (MSSM) analyses it was immediately noted, that a super GIM mechanism is needed in order to satisfy the bounds from flavor changing neutral currents (FCNCs) [30]. Therefore, the mass matrix of the left-handed squarks should be (at least approximately) proportional to the unit matrix, since otherwise flavor off-diagonal entries arise inevitably either in the up or in the down sector due to the SU(2) relation between the left-handed squark mass terms. The idea that non-degenerate squarks can still satisfy the FCNC constraints (K and D mixing) was first discussed in Ref. [31] (an updated analysis can be found in Ref. [32]) in the context of abelian flavor symmetries [33, 34]. In the meantime, there have been a lot of significant improvements both on the theoretical and on the experimental side: The mass difference in the D system was measured and the decay constants and bag factors were calculated to a high precision using lattice methods. A recent analysis of the constraints put on NP by Kaon and D mixing can be found in [35]. In all MSSM analyses the main focus has been on the gluino contributions, while the chargino and neutralino contributions were usually neglected claiming that they are suppressed by a factor of g_2^4/g_s^4 [9–11, 25, 31, 35, 36]. However, it is no longer a good approximation to consider only the gluino contributions in the presence of off-diagonal elements in the LL block of the squark mass matrices because the winos couple to left-handed squarks with g_2. In addition, the gluino contributions suffer from cancellations between the crossed and uncrossed box-diagrams, especially if the gluino is heavier than the squarks. Therefore, the neutralino and chargino contributions can even be dominant if M_2 is light and the gluino is heavier than the squarks. This situation can occur in GUT-motivated scenarios in which the relation $M_2 \approx m_{\tilde{g}}\alpha_2/\alpha_3$ holds. Therefore, we want to update the evaluation of the constraints from K and D mixing with focus on the mass splitting between the first two squark generations taking into account the weak contributions as well.

The squark spectrum is a hot topic concerning bench-mark scenarios for the LHC. It is commonly assumed that the squarks are degenerate at some high scale and that non-degeneracies are introduced via the renormalization group [119, 120]. In such scenarios, the non-degeneracies are proportional to Yukawa couplings and therefore only sizable for the third generation. However, flavor-off-diagonal entries in the squark mass matrix can also lead to non-degenerate squarks which can have an interesting impact on the expected decay and production rate of squarks [121]. In principle, there remains the possibility that squarks have already different masses at some high scale. The question which we want to clarify in this article is which regions in parameter space with non-degenerate squarks are compatible with $D-\overline{D}$ and $K-\overline{K}$ mixing. We are going to discuss this issue in Sec. 7.2 after reviewing $K-\overline{K}$ mixing and $D-\overline{D}$ mixing in Sec. 7.1.

7.1. Meson mixing between the first two generations

Measurements of flavor-changing neutral current (FCNC) processes put strong constraints on new physics at the TeV scale and provide a important guide for model building. In particular $D-\overline{D}$ and $K-\overline{K}$ mixing strongly constrain transitions between the first two generations and combining both is especially powerful to place bounds on new physics [35]. In the down sector FCNCs between the first two generations are probed by the neutral Kaon system, the first observed example of meson- anti-meson mixing. Kaon mixing was already discovered in the early 50th and the CP violation was established in 1964. The up to date experimental values for the mass difference and the CP violating quantity ϵ_K are [67]:

$$\Delta m_K/m_K = (7.01 \pm 0.01) \times 10^{-15}$$
$$\epsilon_K = (2.23 \pm 0.01) \times 10^{-3} \quad (7.1)$$

However, still today, in the age of the B-factories, the long known neutral Kaon system still provides powerful constraints on the flavor structure of any NP model. As we see from (7.1) both the mass difference and the size of the indirect CP violation are tiny and the numbers are in agreement with the standard model (SM) prediction: The SM contribution to the mass difference is small due to a rather precise GIM suppression (the top contribution is suppressed by small CKM elements) and also the CP asymmetry is strongly suppressed because CP violation necessarily involves the tiny CKM combination $V_{td}V_{ts}^*$ related to the third fermion generation. Therefore, Kaon mixing puts very strong bounds on NP scenarios like the MSSM. According to the analysis of Ref. [87] the allowed range in the $C_{M_K}-C_{\epsilon_K}$

plane is rather limited. At 95% confidence level on can roughly expect the NP contribution to the mass difference ΔM_K to be at most of the order of the SM contribution. The NP contribution to ϵ_K is even more restricted. The gluino contribution to $K-\overline{K}$ mixing was in the focus of many analyses [9, 10, 30, 31]. An complete study of the gluino contributions, taking into account the NLO evolution of the Wilson coefficients was done in Ref. [11]. However, neither of these articles considered the electroweak contributions. Only Ref. [122] calculated the chargino contributions but the gluino and neutralino contribution were neglected in this article and the SU(2) relation connecting the up and down squark mass matrices was not used. We return to this point in section III.

In the up sector FCNCs are probed by $D-\overline{D}$ mixing. In contrast to the well established Kaon mixing, it was only discovered recently in 2007 by the BABAR [123] and BELLE [124, 125] collaborations. The current experimental values are [126]:

$$\Delta m_D/m_D = (8.6 \pm 2.1) \times 10^{-15}$$
$$A_\Gamma = (1.2 \pm 2.5) \times 10^{-3} \tag{7.2}$$

Short-distance SM effects are strongly CKM suppressed and the long-distance contributions can only be estimated. Therefore, conservative estimates assume for the SM contribution a range up to the absolute measured value of the mass difference. However, due to the small measured mass difference D mixing still limits NP contributions in a stringent way. Furthermore, a CP phase in the neutral D system can directly be attributed to NP. A first analysis (also including the implications for the MSSM) was done shortly after the experimental discovery [25] and a recent update can be found in Ref. [36]. However, these studies did not consider the electroweak contributions.

In summary, $D-\overline{D}$ and $K-\overline{K}$ mixing restrict FCNC interactions between the first two generations in a stringent way and one should expect the NP contributions to the mass difference to be smaller than the experimental value [35]:

$$\Delta m_{D,K}^{\text{NP}} \leq \Delta m_{D,K}^{\text{exp}} \tag{7.3}$$

CP violation associated with new physics is even more restricted, especially in the down sector:

$$\epsilon_K^{\text{NP}} \leq 0.6 \epsilon_K^{\text{exp}} \tag{7.4}$$

(7.3) and (7.4) summarize in a concise way the allowed range for NP and we will use them to constrain the NP contributions to K and D mixing in Sec. 7.2.

7.2. Constraints on the mass splitting from Kaon mixing and D mixing.

In this section we want to discuss the constraints on the mass splitting between the first two generations of left-handed squark. Due to the $SU(2)$ relation between the left-handed up and down squark mass matrices, $M_{\tilde{u}}^2 = V_{CKM}^\dagger M_{\tilde{d}}^2 V_{CKM}$, in the super-CKM basis, these mass matrices are not independent. The only way to avoid flavor off-diagonal mass insertions in the up and in the down sector simultaneously is to choose $M_{\tilde{q}}^2$ proportional to the unit matrix. This is realized in the naive minimal flavor violating MSSM. In a more general definition of MFV [99] flavor-violation due to NP is postulated to stem solely from the Yukawa sector, resulting in FCNC transitions (which can now also be mediated by gluinos and neutralinos) proportional to products of CKM elements and Yukawa couplings. Therefore, such scenarios allow only sizable deviations from degeneracy with respect to the third generation. However, even though non-degeneracies with the third generation induce additional CP violation associated with V_{ub} we find that this mass splitting effectively cannot be constrained. This finding is in agreement with Ref. [127] A bit more general notion of MFV could be defined by stating that all flavor change should be induced by CKM elements. This definition would also cover the case with a diagonal squark mass matrix in one sector (either the up or the down sector) but with off-diagonal elements, introduced by the $SU(2)$ relation, in the other sector. This setup corresponds to an exact alignment of the squark mass term $m_{\tilde{q}}^2$ with the product of Yukawa matrices $Y_u^\dagger Y_u$ (or with $Y_d^\dagger Y_d$ in the case of a diagonal down squark mass matrix).

The obvious way how off-diagonal elements of the squark mass matrices enter meson mixing is via squark-gluino diagrams. These contributions are commonly expected to be dominant since they involve the strong coupling constant. Also in our case under study, with flavor-violating LL elements, the gluino diagrams were assumed to be the most important SUSY contributions to the Wilson coefficient C_1 of the $\Delta F = 2$ effective Hamiltonian $H_{eff}^{\Delta F=2} = \sum_{i=1}^{5} C_i O_i + \sum_{i=1}^{3} \tilde{C}_i \tilde{O}_i$ [9–11, 25, 31, 35, 36]:

$$C_1^{\tilde{g}\tilde{g}} = -\frac{g_s^4}{16\pi^2} \sum_{s,t=1}^{6} \left[\frac{11}{36} D_2\left(m_{\tilde{q}_s}^2, m_{\tilde{q}_t}^2, m_{\tilde{g}}^2, m_{\tilde{g}}^2\right) + \frac{1}{9} m_{\tilde{g}}^2 D_0\left(m_{\tilde{q}_s}^2, m_{\tilde{q}_t}^2, m_{\tilde{g}}^2, m_{\tilde{g}}^2\right) \right] V_{s\,12}^{q\,LL} V_{t\,12}^{q\,LL} \tag{7.5}$$

Our conventions for the loop-functions are given in the appendix and the matrices in flavor space $V_{s\,12}^{q\,LL}$ are defined in equation (5.6). However, if we have flavor-changing LL elements it is no longer possible to concentrate on the gluino contributions for four reasons:

- The gluino contributions suffer from cancellations between the boxes with crossed and uncrossed gluino lines corresponding to the two terms in the square brackets in (7.5). The crossed box diagrams occur since the gluino is a Majorana particle. This cancellation occurs approximately in the region where $m_{\tilde{g}} \approx 1.5\, m_{\tilde{q}}$

- In the SU(2) limit with unbroken SUSY the winos couple directly to left-handed particles with the weak coupling constant g_2. Therefore, flavor-changing LL elements can contribute without involving small left-right or gaugino mixing angles.

- Since charginos are Dirac fermions, there are no cancellations between different diagrams at the one-loop order.

- The wino mass M_2 is often assumed to be much lighter than the gluino mass. In most GUT models the relation $M_2 \approx m_{\tilde{g}}\alpha_2/\alpha_3$ holds. Since the loop function is always dominated by the heaviest mass, one can expect large chargino and neutralino contributions if the squarks masses are similar to the lighter chargino masses.

Therefore, we have to take into account the weak (and the mixed weak-strong) contributions to C_1:

$$C_1^{\tilde{\chi}^0\tilde{\chi}^0} = -\frac{1}{128\pi^2}\frac{g_2^4}{4}\sum_{s,t=1}^{6} \left(D_2\left(m_{\tilde{q}_s}^2, m_{\tilde{q}_t}^2, M_2^2, M_2^2\right)\right.$$
$$\left. + 2M_2^2 D_0\left(m_{\tilde{q}_s}^2, m_{\tilde{q}_t}^2, M_2^2, M_2^2\right)\right) V_{s\,12}^{q\,LL} V_{t\,12}^{q\,LL}$$

$$C_1^{\tilde{g}\tilde{\chi}^0} = -\frac{1}{16\pi^2}\frac{g_s^2 g_2^2}{2}\sum_{s,t=1}^{6}\left(\tfrac{1}{6}D_2\left(m_{\tilde{q}_s}^2, m_{\tilde{q}_t}^2, m_{\tilde{g}}^2, M_2^2\right)\right. \quad (7.6)$$
$$\left. + \tfrac{1}{3}m_{\tilde{g}}M_2 D_0\left(m_{\tilde{q}_s}^2, m_{\tilde{q}_t}^2, m_{\tilde{g}}^2, M_2^2\right)\right) V_{s\,12}^{q\,LL} V_{t\,12}^{q\,LL}$$

$$C_1^{\tilde{\chi}^+\tilde{\chi}^+} = -\frac{g_2^4}{128\pi^2}\sum_{s,t=1}^{6} D_2\left(m_{\tilde{q}_s}^2, m_{\tilde{q}_t}^2, M_2^2, M_2^2\right) V_{s\,12}^{q\,LL} V_{t\,12}^{q\,LL}$$

In (7.6) we have set all Yukawa couplings to zero and neglected small chargino and neutralino mixing. Due to the small Yukawa couplings of the first two generations and the suppressed bino-wino mixing the only sizable contribution of both the gluino and the electroweak diagrams is to the same operator $O_1 = \bar{s}\gamma^\mu P_L d \otimes \bar{s}\gamma_\mu P_L d$ as the SM contribution. Note that in all contributions the same combination of mixing matrices enters, since the CKM matrices in the chargino vertex cancels with the ones in the squark mass matrix. Ref. [128] calculated all Wilson coefficients contributing to $\Delta F = 2$ processes in the MSSM and Ref. [27] included also the chargino and neutralino contributions into their numerical

7.2 Constraints on the mass splitting from Kaon mixing and D mixing.

analysis. However, the main focus of Ref. [27] is not on the mass-splitting between the first two squark generations and the importance of the different contributions is not apparent from the scatter plots used in their analysis.

In Fig. 7.1 we show the size of the different contributions to C_1 as a function of the gluino mass. We have normalized all coefficients to $C_1^{\tilde{\chi}^+\tilde{\chi}^+}$ since only one box diagram contributes to it and therefore the coefficient depends only on one loop-function which is strictly negative. Note that for heavy gluino masses always the chargino and in some cases the mixed gluino-neutralino contribution are dominant.

As stated before, SU(2) symmetry links a mass splitting in the up (down) sector to flavor-changing LL elements in the down (up) sector. So, if one assumes a "next-to minimal" setup in which one mass matrix is diagonal, one has to specify if this is the up or the down squark mass matrix. If the down (up) squark mass matrix is diagonal, which implies that it is aligned to $Y_d^\dagger Y_d$ ($Y_u^\dagger Y_u$), one has contributions to $D-\bar{D}$ ($K-\bar{K}$) mixing.

Assuming a diagonal up-squark (down-squark) mass matrix, the regions in the $m_{\tilde{u}_1}$-$m_{\tilde{g}}$ plane compatible with $K-\bar{K}$ mixing ($D-\bar{D}$ mixing) are shown in Fig. 7.2. Note that there are large regions in parameter space with non-degenerate squark still allowed by $K-\bar{K}$ ($D-\bar{D}$) mixing due to the cancellations between the different contributions shown in Fig. 7.1. However, departing from an exact alignment with either $Y_u^\dagger Y_u$ or $Y_d^\dagger Y_d$ there are points in parameter space which allow for an even larger mass splitting [35] due to an additional off-diagonal element in the squark mass matrix. If this element is real one can choose an appropriate value which maximizes the allowed mass splitting [1]. Nevertheless, this additional off-diagonal element now present in both sectors due to the SU(2) relation could also carry a phase additional to the CKM matrix. If this phase is maximal one obtains the minimally allowed range for the mass splitting due to the severe constraint from ϵ_K. These minimally and maximally allowed regions for the mass splittings are also shown in Fig. 7.2.

We have seen that due to the cancellations between the different diagrams contributing to D–$\bar{D}$ and K–$\bar{K}$ mixing there are large allowed regions in parameter space where the squarks are not degenerate (a mass splitting of 100% and more is well possible). This has also interesting consequences for the LHC: While most benchmark scenarios assume degenerate squark masses [119, 120] non-degenerate masses can have interesting consequences on the branching ratios [121]. The conclusion we can draw from Fig. 7.2 is that there are regions in parameter space, allowed by $K-\bar{K}$ and $D-\bar{D}$ mixing, with very different masses for

---
[1]We thank Gilad Perez for bringing this to our attention.

the first two squark generations. Therefore, FCNC processes alone do not require the soft-SUSY breaking parameter $M_{\tilde{q}}^2$ to be proportional to the unit matrix at some high scale. This implicates that there is more allowed parameter space for models with abelian flavor symmetries than without the inclusion of the electroweak contributions to D–$\overline{D}$ and K–$\overline{K}$ mixing.

7.2 Constraints on the mass splitting from Kaon mixing and D mixing. 79

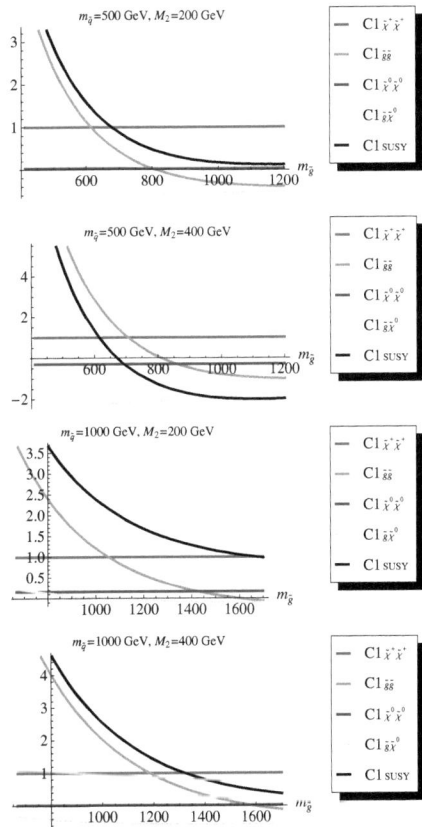

Figure 7.1: Size of the real part of Wilson coefficients (see (7.5) and (7.6)) contributing to $D-\overline{D}$ or $K-\overline{K}$ mixing normalized to the chargino contribution as a function of $m_{\tilde{g}}$ for different values of $m_{\tilde{q}}$ and M_2 assuming a small non-zero (real) off-diagonal element $\delta_{12}^{q\,LL}$. $C_{1\text{SUSY}}$ is the sum of all Wilson coefficients contributing in addition to the SM one. The relative size of the coefficients remains unchanged also in the case of complex elements $\delta_{12}^{q\,LL}$.

80   7. LL Mass Insertions and constraints from K and D mixing

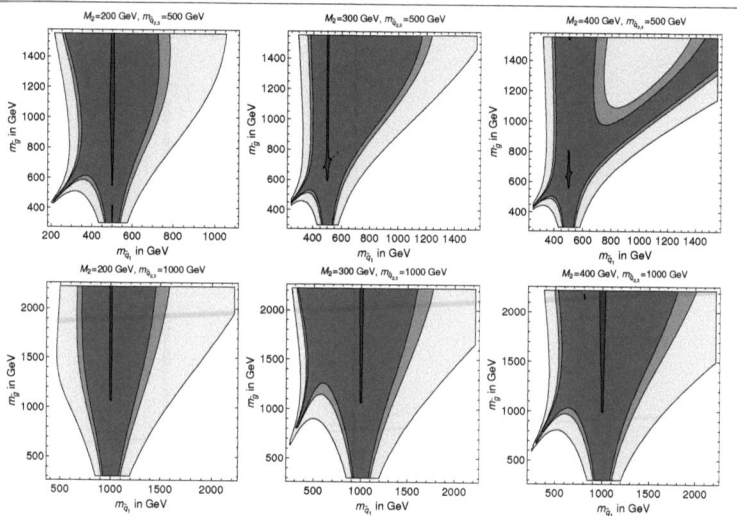

Figure 7.2: Allowed regions according to (7.3) and (7.4) in the $m_{\tilde{q}_1} - m_{\tilde{g}}$ plane with $m_{\tilde{q}_{2,3}} = 500, 1000$ GeV for different values of M_2. Yellow (lightest) corresponds to the maximally allowed mass splitting assuming an intermediate alignment of $m_{\tilde{q}}^2$ with $Y_u^\dagger Y_u$ and $Y_d^\dagger Y_d$ [35]. The green (red) region is the allowed range assuming an diagonal up (down) squark mass matrix. The blue (darkest) area is the minimal region allowed for the mass splitting between the left-handed squarks, which corresponds to a scenario with equal diagonal entries in the down squark mass matrix but with an off-diagonal element carrying a maximal phase.

8. RIGHT-HANDED W COUPLING AND ITS EFFECTS ON V_{ub} AND V_{cb}

In the standard model (SM) with its gauge group $SU(3)_C \times SU(2)_L \times U(1)$ the tree-level W coupling has a pure $V - A$ structure meaning that all charged currents are left-handed. Right-handed charged currents were first studied in the context of left-right symmetric models [37] which enlarge the gauge group by an additional $SU(2)_R$ symmetry between right-handed doublets. In these models new right-handed gauge bosons W_R, Z_R appear and the physical SM-like W-boson has a dominant left-handed component with a small admixture of W_R. The latter will generically lead to small right-handed couplings to both quarks and leptons. The right-handed mass scale inferred from today's knowledge on neutrino masses is so large that all right-handed gauge couplings are undetectable. Most of these couplings are further experimentally strongly constrained [67]. A different source of right-handed couplings of quarks to the W-boson can be loop effects, which generate a dimension-6 quark-quark-W vertex. In this case no right-handed lepton couplings occur, as long as the neutrinos are assumed left-handed. A generic analysis of such higher-dimensional right-handed couplings has been studied in Ref. [38] aiming at a better understanding of $K \to \pi\mu\nu$ data. The general effect of left- and right-handed anomalous couplings of the W to charm was studies in Ref. [39]. The authors conclude that only the real part of the right-handed charm-bottom coupling can be sizable. The coupling of the W to up has been studied in [40].

We will investigate the effect of a right-handed W-coupling on the extraction of $|V_{ub}|$ and $|V_{cb}|$ in section II and show that current tensions between SM and data can be removed. In section III we will calculate the loop-corrected W-coupling in the generic Minimal Supersymmetric Standard Model (MSSM). We find that the right-handed W-coupling can be as large as 20% and brings the different determinations of $|V_{ub}|$ into perfect agreement. The effect on $|V_{cb}|$ is at most around 2%, which alleviates the tension studied in the next section.

8.1. Right-handed W couplings

An appropriate framework for our analysis is an effective Lagrangian. Following the notation of Ref. [129], we write

$$\mathcal{L} = \mathcal{L}_{\text{SM}} + \frac{1}{\Lambda}\sum_i C_i^{(5)} Q_i^{(5)} + \frac{1}{\Lambda^2}\sum_i C_i^{(6)} Q_i^{(6)} + \mathcal{O}\left(\frac{1}{\Lambda^3}\right), \qquad (8.1)$$

here $\mathcal{L}_{\text{SM}}$ is the standard model (SM) Lagrangian, while $Q_i^{(n)}$ stand for dimension-n operators built out of the SM fields and are invariant under the SM gauge symmetries. Such an effective theory approach is appropriate for any SM extension in which all new particles are sufficiently heavy ($M_{\text{new}} \sim \Lambda \gg m_t$). As long as only processes with momentum scales $\mu \ll \Lambda$ are considered, all heavy degrees of freedom can be eliminated [58], leading to the effective theory defined in (8.1). The operators $Q_i^{(5)}$ and $Q_i^{(6)}$ have been completely classified in Ref. [130]. Since $Q_i^{(5)}$ involve no quark fields, they are not needed for our further discussion, and we skip the superscripts "(6)" at the dimension-six operators and the associated Wilson coefficients C_i. In this article, we need the following dimension-six operator describing anomalous right-handed W-couplings to quarks:

$$Q_{RR} = \bar{u}_f \gamma^\mu P_R d_i \left(\tilde{\phi}^\dagger i D_\mu \phi\right) + \text{h.c.} \qquad (8.2)$$

where ϕ denotes the Higgs doublet and $\tilde{\phi} = i\tau^2 \phi^*$. The Feynman rule for the W-u_f-d_i interaction vertex,

$$\frac{-ig_2\gamma^\mu}{\sqrt{2}}\left(V_{fi}^L P_L + V_{fi}^R P_R\right), \qquad (8.3)$$

is found by combining the usual SM interaction with the extra contributions that are obtained by setting the Higgs field in (8.2) to its vacuum expectation value. In (8.3) V_{fi}^L and V_{fi}^R denote elements of the effective CKM matrices, which are not necessarily unitary. V_{fi}^R is related to the Wilson coefficient in (8.1) via $V_{fi}^R = \frac{C_{RR}}{2\sqrt{2}G_F \Lambda^2}$. V_{fi}^L receives contributions from the tree-level CKM matrix and the LL analogue of Q_{RR} in (8.2).

Right-handed couplings to light quarks have been studied in Ref. [38] and to charm (up) quarks in Ref. [39] (Ref. [40]). Ref. [131] examines such couplings in inclusive b→c transitions. In Ref. [129] it was pointed out that very strong constraints can be obtained on V_{tb}^R from $b \to s\gamma$, because the usual helicity suppression factor of m_b/M_W is absent in the right-handed contribution. By the same argument V_{ts}^R (or V_{td}^R if one considers $b \to d\gamma$) is tightly constrained. Large effects concerning transitions between the first two generations are unlikely, because V_{us}^L and V_{cd}^L are larger than other off-diagonal CKM elements. Further deviations from Minimal Flavour Violation (as defined in [97]), i.e. deviations from

8.1 Right-handed W couplings

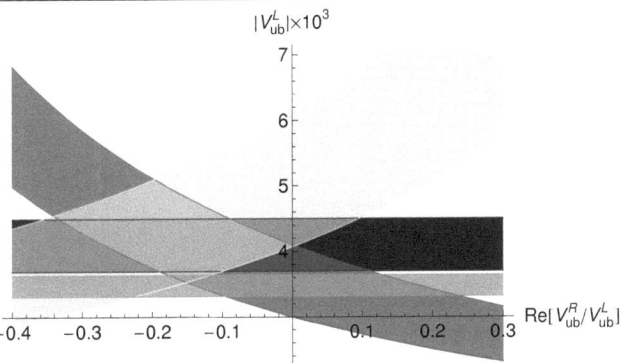

Figure 8.1: $\left|V_{ub}^L\right|$ as a function of $\text{Re}\left[V_{ub}^R/V_{ub}^L\right]$ extracted from different processes. Blue(darkest): inclusive decays. Red(gray): $B \to \pi l\nu$. Yellow(lightest gray): $B \to \tau\nu$. Green(light gray): V_{ub}^L determined from CKM unitarity.

Yukawa-driven flavour transitions, are unlikely in the first two generations, but plausible with respect to transitions involving the third generation. We therefore focus our attention on the remaining two elements V_{ub}^R and V_{cb}^R.

8.1.1. Determination of V_{ub}^L and V_{cb}^L

The experimental determination of $|V_{ub}|$ and $|V_{cb}|$ from both inclusive and exclusive B decays is a mature field by now [67]. E.g. the form factors needed for $B \to \pi l\nu$ are known to 12% accuracy [132]. More recently, also the leptonic decay $B \to \tau\nu_\tau$ is studied in the context of V_{ub}. To discuss the impact of right-handed currents we denote the CKM element extracted from data with SM formula by V_{qb}, where $q = u$ or $q = c$. If the matrix element of a considered exclusive process is proportional to the vector current, V_{qb}^L and V_{qb}^R enter with the same sign and the "true" value of V_{qb}^L in the presence of V_{qb}^R is given by:

$$V_{qb}^L = V_{qb} - V_{qb}^R \qquad (8.4)$$

For processes proportional to the axial-vector current V_{qb}^R enters with the opposite sign as V_{qb}^L, so that

$$V_{qb}^L = V_{qb} + V_{qb}^R. \qquad (8.5)$$

In inclusive decays the interference term between the left-handed and right-handed contributions is suppressed by a factor of m_q/m_b, so that it is irrelevant in the case of V_{ub} and somewhat suppressed in the case of V_{cb}. The remaining dependence on V_{qb}^R is quadratic and therefore negligible.

Starting with $|V_{ub}|$, we note that the determinations from inclusive and exclusive semileptonic decays agree within their errors, but the agreement is not perfect [67, 86]. The analysis of $B \to \tau\nu$ is affected by the uncertainty in the decay constant f_B. Within errors the three determinations of $|V_{ub}|$ are compatible, as one can read off from Fig. 8.1. The picture looks very different once the information from a global fit to the unitarity triangle (UT) is included: As pointed out first by the CKMFitter group, the measured value of $B \to \tau\nu$ suffers from a tension with the SM of 2.4–2.7σ [86]. First, the global UT fit gives a much smaller error on $|V_{ub}|$ (as a consequence of the well-measured UT angle β); the corresponding value is also shown in Fig. 8.1. Second, the data on B_d–$\bar{B}_d$ mixing exclude very large values for f_B, which in turn cut out the lower part of the yellow (light gray) region in Fig. 8.1. Essentially we realize from Fig. 8.1 that we can remove this tension while simultaneously bringing the determinations of $|V_{ub}|$ from inclusive and exclusive semileptonic decays into even better agreement. For this the right-handed component must be around $\mathrm{Re}\,(V_{ub}^R/V_{ub}^L) \approx -0.15$. Since new physics may as well affect the other quantities entering the UT, a more quantitative statement requires the consideration of a definite model.

Next we turn to $|V_{cb}|$: The relative uncertainties in the exclusive decays $B \to D^*l\nu$ and $B \to Dl\nu$ and in the inclusive $B \to X_c\ell\nu$ analyses are much smaller than in the $b \to u$ decays considered above. Note that $B \to Dl\nu$ only involves the vector current so that (8.4) applies. $B \to D^*l\nu$ receives contributions from both vector and axial vector currents, but the contribution from the vector current is suppressed in the kinematic endpoint region used for the extraction of $|V_{cb}|$. Therefore (8.5) applies to $B \to D^*l\nu$. The impact of a right-handed current on $B \to X_c\ell\nu$ has been calculated in Ref. [131]. Fig. 8.2 shows that the agreement among the three values of $|V_{cb}|$ obtained from these decay modes is not totally satisfactory within the SM. One further realizes that we can reduce the discrepancy to less than 1σ if a right-handed coupling in the range $0.03 \leq \mathrm{Re}\,[V_{cb}^R/V_{cb}^L] \leq 0.06$ is present.

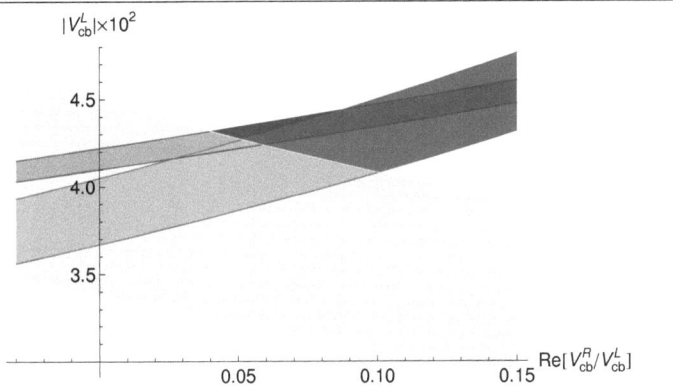

Figure 8.2: $\left|V_{cb}^L\right|$ as a function of $\text{Re}\left[V_{cb}^R/V_{cb}^L\right]$ extracted from different processes. Blue(darkest): inclusive decays. Red(gray): $B \to D^*l\nu$. Yellow(light gray): $B \to Dl\nu$.

8.2. SQCD corrections to the quark-quark-W vertex

In section 3.4 the impact of chirally-enhanced non-decoupling self-energies on of the quark-quark-W vertex has been discussed. These corrections are unitary and therefore result in a renormalization of the CKM matrix as required by the decoupling theorem [58]. In this section we calculate the leading contributions to the quark-quark-W vertex which decouple for $M_{\text{SUSY}} \to \infty$.

The self energies lead to a flavor-valued wave-function renormalization $\Delta U^{q\,L,R}_{fi}$ (see (3.4)) for all external left- and right-handed fields. It is useful to decompose these factors further into an unphysical anti-Hermitian part $\Delta U^{q\,L\,A}_{fi}$, which can be absorbed into the renormalization of the CKM matrix, and a Hermitian part $\Delta U^{q\,L\,H}_{fi}$, which can constitute a physical effect appearing as a deviation from CKM unitarity:

$$\Delta U^{q\,L,R\,H}_{fi} = \Sigma^{q\,LL,RR}_{fi}/2. \tag{8.6}$$

Neglecting external momenta, the genuine vertex-correction originating from a squark-

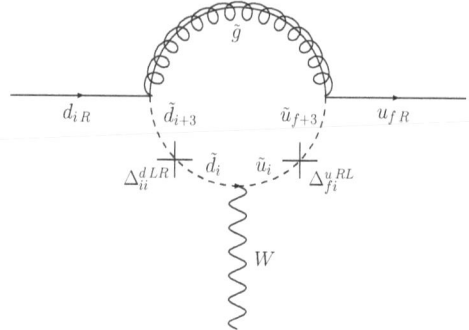

Figure 8.3: Feynman diagram which induces the effective right-handed W coupling of a down-type quark of flavor i to a up-type quark of flavor f. The crosses stand for the flavor and chirality changes needed to generate the coupling.

gluino loop is given by

$$-i\Lambda^{W\tilde{g}}_{u_f d_i} = \frac{g_2}{\sqrt{2}} \frac{i\alpha_s}{3\pi} \gamma^\mu \sum_{s,t=1}^{6} \sum_{j,k=1}^{3} \left(W^{\tilde{u}}_{fs} W^{\tilde{u}*}_{ks} V^L_{kj} W^{\tilde{d}}_{jt} W^{\tilde{d}*}_{it} P_L + W^{\tilde{u}}_{f+3,s} W^{\tilde{u}*}_{ks} V^L_{kj} W^{\tilde{d}}_{jt} W^{\tilde{d}*}_{i+3,t} P_R \right)$$
$$\times C_2 \left(m^2_{\tilde{u}_s}, m^2_{\tilde{d}_t}, m^2_{\tilde{g}} \right)$$
(8.7)

The part proportional to P_L in (8.7) cancels with the anti-Hermitian part of the wave-function renormalization due to the SU(2) relation between the left-handed up and down squarks for $M_{\text{SUSY}} \to \infty$ according to the decoupling theorem [58]. Since the loop functions depend only weakly on M_{SUSY}, the cancellation is very efficient, even for light squarks around 300 GeV. Therefore, the unitarity of the CKM matrix is conserved with very high accuracy. A right-handed coupling of quarks to the W boson is induced by the diagram in Fig. 8.3 if left-right mixing of squarks is present. The effective coupling corresponds to Q_{RR} in (8.2) and vanishes in the decoupling limit. There is no wave-function renormalization of right-handed quarks which can be applied to the W vertex, therefore no gauge cancellations occur. We show the relative size of the right-handed coupling involving u,c and b in Fig. 8.4. Note that the mass insertion $\delta^{u\,RL}_{13,23}$ are not affected by the fine-tuning argument imposed in chapter 4 nor severely restricted by FCNC processes [73]. Therefore, the size of the induced couplings V^R_{ub} (V^R_{cb}) can be large enough to explain (attenuate) the apparent discrepancies

8.2 SQCD corrections to the quark-quark-W vertex

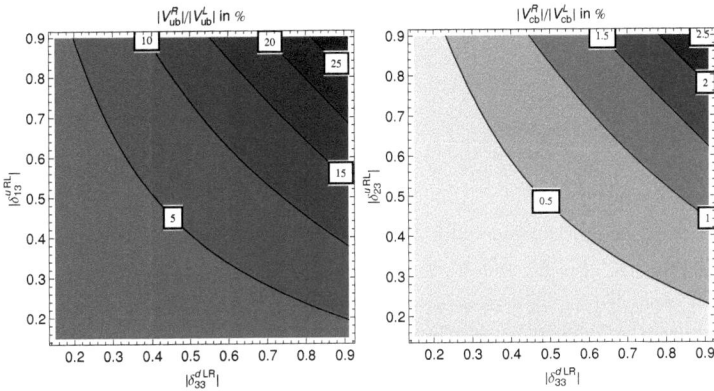

Figure 8.4: Left: Relative strength of the induced right-handed coupling $|V_{ub}^R|$ with respect to $|V_{ub}^L|$ for $M_{\text{SUSY}} = 1\,\text{TeV}$. $|V_{ub}^L|$ is determined from CKM unitarity. Right: Same as the left figure for V_{cb}^R.

among the various determinations of $|V_{ub}|$ ($|V_{cb}|$). Nevertheless, if $\delta_{13,23}^{u\,RL}$ is large single-top production is enhanced which can be observed at the LHC [74]. In principle also charged Higgs contributions to $B \to \tau\nu$ have to be considered in the MSSM. However, these contributions are only important in the special case in which both $\tan(\beta)$ is large and the charged Higgs is light. Furthermore, a charged Higgs always interferes destructively with the SM, making the discrepancy between the different determinations of V_{ub} even bigger.

9. Conclusions

In this thesis we have examined the effects of non-minimal sources of flavor violation in the MSSM.

First we computed the finite renormalization of fermion masses and mixing angles, taking into account the leading two-loop effects. These corrections are not only important in order to obtain a unitary CKM matrix, they are also numerically significant for light fermion masses. As another application of these results, we have derived supersymmetric loop corrections to the couplings of charged Higgs bosons and charginos to quarks and squarks. In these couplings the squark-gluino loops which renormalize the CKM elements are physical and can have a significant numerical impact because of their chiral enhancement. We have further pointed out that the calculated flavor-changing self-energies can have observable effects in the W-mediated production or decay of the top quark, with the SUSY effects decoupling as m_t^2/M_{SUSY}^2 for $M_{\text{SUSY}} \to \infty$.

According to 't Hooft's naturalness principle, the smallness of a quantity is linked to a symmetry that is restored if the quantity is zero. This argument applies to the small Yukawa couplings of the first two generations (as well as to the small CKM elements involving the third generation) since a flavor symmetry is gained if the Yukawa couplings are set to zero. We use 't Hooft's naturalness criterion to constrain the chirality-changing mass insertion $\delta_{IJ}^{u,d,\ell\, LR}$ from the mass and CKM renormalization. All constraints given in this context are non-decoupling. This means they do not vanish in the limit of infinitely heavy SUSY masses unlike the bounds from FCNC processes. Therefore, our constraints are always stronger than the FCNC constraints for sufficiently heavy SUSY (and Higgs) masses. The NLO corrections also allow us to constrain the product $\delta_{13}^{f\, LR}\delta_{31}^{f\, LR}$ (and $\delta_{23}^{d\, LR}\delta_{32}^{d\, LR}$) which is important because the elements $\delta_{13,23}^{u\, RL}$ were unconstrained.

In the analysis of FCNC processes in the generic MSSM we have pointed out that they receive chirally enhanced two-loop or three-loop corrections which can numerically dominate over the usual one-loop diagrams. The chirally enhanced contributions involve a flavor change in a self-energy sub-diagram attached to an external leg of the diagram. These ef-

9. Conclusions

fects can be absorbed into a finite renormalization of the squark-quark-gluino vertex. Our new effects vanish if the squark masses are degenerate. Their relative importance with respect to the LO diagrams is larger for heavier squarks.

In our phenomenological study of FCNC processes we first addressed $B \to X_s\gamma$. In this process our new effects are only relevant if $|\mu|\tan\beta$ is large. We presented new bounds on the four quantities $\delta_{23}^{d\,LL}$, $\delta_{23}^{d\,LR}$, $\delta_{23}^{d\,RL}$, and $\delta_{23}^{d\,RR}$ which parametrize the off-diagonal elements of the down-squark mass matrix linking strange and bottom squarks for large $\tan\beta$. These bounds are depicted in Fig. 5.4 for the case of real MSSM parameters. As a general pattern we find that the chirally enhanced contributions decrease the size of the SUSY contribution to $\text{Br}\,[B \to X_s\gamma]$ if μ is positive. Conversely, the chirally enhanced two-loop contributions increase the SUSY contribution to $\delta_{23}^{d\,AB}$ for $\mu < 0$. That is, for positive values of μ, which are preferred by the anomalous magnetic moment of the muon, the bounds on $\delta_{23}^{d\,AB}$ become weaker.

As a second application we studied the chirally enhanced effects in B_d, B_s, and K mixing. Using the data on the mass differences ΔM_d and ΔM_s and on CP asymmetries we find new constraints on the complex $\delta_{13}^{d\,LR}$, $\delta_{23}^{d\,LR}$, $\delta_{13}^{d\,RL}$, and $\delta_{23}^{d\,RL}$ elements (see Fig. 5.7). In most of the parameter space the constraints become much stronger compared to the LO analysis if the sbottom mass differs sizably from the squark masses of the first two generations, irrespective of the size of $\tan\beta$. K–$\overline{K}$ mixing is even more sensitive to the chirally enhanced self-energies, provided there is a non-zero mass splitting among the squarks of the first two generations. As illustrated in Fig. 5.9 already mass splittings in the sub-percent range strengthen the bounds on $\delta_{12}^{d\,LR}$ and $\delta_{12}^{d\,RL}$ severely.

Radiative generation of light fermion masses is a pleasant scenario. Within the MSSM this approach can solve the SUSY CP and the SUSY flavor problem. We studied this model and its phenomenological consequences in chapter 6. Keeping the third generation fermion Yukawa coupling, the CKM matrix can either be induced in the up or in the down sector (in principle also a mixed scenario is possible, however, we did not further investigate this possibility). If the CKM matrix is generated in the up-sector Kaon mixing severely constraints the allowed values of $m_{\tilde{g}}$ and $m_{\tilde{q}}$ (see Fig. 6.1 b)). However, the rare Kaon decay $K \to \pi\nu\nu$ can still receive sizable contributions. If the other possibility is realized and the CKM matrix is generated in the down sector, $b \to s\gamma$ restricts the allowed rage for the SUSY masses. Taking into account the chirally enhanced correction discussed in chapter 5 our results are shown in Fig. 6.1 a). The implications for the lepton sector are even more significant. The anomalous magnetic moment of the muon restricts the allowed range of the lighter smuon mass to be roughly between 1 TeV and 3 TeV (see Fig. 6.4).

In chapter 7 we have examined the constraints on the mass splitting between the first two generations of left-handed squarks from $K-\overline{K}$ and $D-\overline{D}$ mixing by considering the gluino and the electroweak contribution. While nearly all previous analysis were restricted to the gluino contributions to $K-\overline{K}$ and $D-\overline{D}$ mixing in the case of non-minimal flavor violation [10, 13–15, 29, 35, 39] Ref. [31] included (but only numerically) the electroweak effects. However, the main focus of Ref. [31] is not on the mass splitting between the squarks and the importance of the different contributions is not apparent from the scatter plots shown in their article. In our analysis we have examined in detail the size of the different contributions (neutralino, neutralino-gluino, gluino and chargino boxes) to $D-\overline{D}$ and $K-\overline{K}$ mixing in the presence of flavor off-diagonal mass-insertions in the LL sector of the squark mass matrices. It is found that gluino contributions suffer from a cancellation between the crossed and the uncrossed boxes for $m_{\tilde{g}} \approx 1.5\, m_{\tilde{q}}$. In addition, winos couple directly to left-handed squark fields (without involving small gaugino or left-right mixing) and their contribution is not affected by such a cancellation. Therefore, we conclude that the (usually neglected) contributions from chargino, neutralino and mixed neutralino-gluino diagrams can be of the same order as (or even dominant over) the gluino contribution especially if $M_2 \approx m_{\tilde{q}} < m_{\tilde{g}}$.

In the analysis of the allowed mass splitting between the first two generations we focused on the "minimal case" in which the up (down) squark mass matrix is diagonal in the super-CKM basis, but not proportional to the unit matrix. In this case flavor off-diagonal elements in the down (up) sector are induced via the SU(2) relation and therefore contribute to $K-\overline{K}\,(D-\overline{D}\,)$ mixing. It is found that the constraints on the mass splitting are strong for light gluino masses. However, if the gluino is heavier than the squarks there are large regions in parameter space, allowed by $K-\overline{K}\,(D-\overline{D}\,)$ mixing, with highly non-degenerate squark masses. This has interesting consequences both for LHC benchmark scenarios (which usually assume degenerate squarks for the first two generations) and for models with abelian flavor symmetries (which predict non-degenerate squark masses for the first two generation) because $K-\overline{K}$ and $D-\overline{D}$ mixing cannot exclude non-degenerate squark masses of the first two generations.

Chapter 8 is devoted to the study of an effective right-handed coupling of quarks to the W boson and its effects on the determination of $|V_{ub}|$ and $|V_{cb}|$ from different decay modes. In both cases a right-handed coupling can improve the agreement among these determinations (Figs. 8.1 and 8.2). In particular, one can simultaneously remove the disturbing

9. Conclusions

problem with $B \to \tau \nu_\tau$ [89] and improve the agreement among inclusive and exclusive determinations of $|V_{ub}|$. Second, we have shown that a loop-induced right-handed coupling is generated within the MSSM if left-right mixing of squarks is present. This coupling has the right size needed to resolve the tensions in $|V_{ub}|$. Such a scenario involves a large left-right mixing between sbottoms (as present in e.g. the popular large-tan β scenarios) and a large A_{31}^u-term which enhances single-top production, making it observable at the LHC. If $\delta_{13}^{u\,RL} \approx 0.6$ a 95% CL signal can already be detected with 50 inverse femto-barn [77]. In $b \to c$ transitions the loop-induced supersymmetric right-handed coupling can alleviate, but cannot fully remove, the discrepancies among the three methods to determine $|V_{cb}|$. To probe $b \to u$ transitions we propose to look for right-handed couplings in the differential decay distributions of $B \to \rho \ell \nu_\ell$. The smaller right-handed component in $b \to c$ transitions can be probably better studied in $B \to X_c \ell \nu$ [131] than in $B \to D^* \ell \nu$ decays, because a theoretical control of form factors to percent accuracy is challenging.

10. Appendix

10.1. Feynman rules

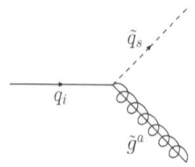

$$i\left[\Gamma_{q_i\tilde{q}_s}^{\tilde{g}L}P_L + \Gamma_{q_i\tilde{q}_s}^{\tilde{g}R}P_R\right] = -i\sqrt{2}g_s T^a \left(W_{is}^{\tilde{q}*}P_L - W_{i+3,s}^{\tilde{q}*}P_R\right) \quad (10.1)$$

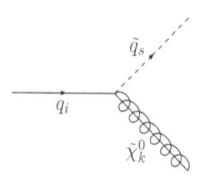

$$i\left[\Gamma_{q_i\tilde{q}_s}^{\tilde{\chi}_k^0 L}P_L + \Gamma_{q_i\tilde{q}_s}^{\tilde{\chi}_k^0 R}P_R\right] \text{ with}$$

$$\Gamma_{d_i\tilde{d}_s}^{\tilde{\chi}_k^0 L} = \sqrt{2}g_2\left(\tfrac{1}{2}Z_{k2}^{\tilde{\chi}^0*} - \tfrac{1}{6}\tan(\theta_W)Z_{k1}^{\tilde{\chi}^0*}\right)W_{is}^{\tilde{d}*} - Y^{d_i}Z_{k3}^{\tilde{\chi}^0}W_{i+3,s}^{\tilde{d}*}$$

$$\Gamma_{d_i\tilde{d}_s}^{\tilde{\chi}_k^0 R} = \tfrac{-\sqrt{2}}{3}g_2\tan(\theta_W)Z_{k1}^{\tilde{\chi}^0}W_{i+3,s}^{\tilde{d}*} - Y^{d_i}Z_{k3}^{\tilde{\chi}^0}W_{is}^{\tilde{d}*}$$

$$\Gamma_{u_i\tilde{u}_s}^{\tilde{\chi}_k^0 L} = -\sqrt{2}g_2\left(\tfrac{1}{2}Z_{k2}^{\tilde{\chi}^0*} + \tfrac{1}{6}\tan(\theta_W)Z_{k1}^{\tilde{\chi}^0*}\right)W_{is}^{\tilde{u}*} - Y^{u_i}Z_{k4}^{\tilde{\chi}^0}W_{i+3,s}^{\tilde{u}*}$$

$$\Gamma_{u_i\tilde{u}_s}^{\tilde{\chi}_k^0 R} = \tfrac{2\sqrt{2}}{3}g_2\tan(\theta_W)Z_{k1}^{\tilde{\chi}^0}W_{i+3,s}^{\tilde{u}*} - Y^{u_i}Z_{k4}^{\tilde{\chi}^0}W_{is}^{\tilde{u}*}$$

$$(10.2)$$

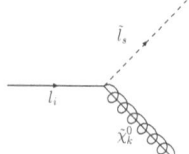

$$i\left[\Gamma_{\ell_i\tilde{\ell}_s}^{\tilde{\chi}_k^0 L}P_L + \Gamma_{\ell_i\tilde{\ell}_s}^{\tilde{\chi}_k^0 R}P_R\right] \text{ with}$$

$$\Gamma_{\ell_i\tilde{\ell}_s}^{\tilde{\chi}_k^0 L} = \tfrac{g_2}{\sqrt{2}}\left(Z_{k2}^{\tilde{\chi}^0*} + \tan(\theta_W)Z_{k1}^{\tilde{\chi}^0*}\right)W_{is}^{\tilde{\ell}*} - Y^{\ell_i}Z_{k3}^{\tilde{\chi}^0*}W_{i+3,s}^{\tilde{\ell}*} \quad (10.3)$$

$$\Gamma_{\ell_i\tilde{\ell}_s}^{\tilde{\chi}_k^0 R} = -\sqrt{2}g_2\tan(\theta_W)Z_{k1}^{\tilde{\chi}^0}W_{i+3,s}^{\tilde{\ell}*} - Y^{\ell_i}Z_{k3}^{\tilde{\chi}^0}W_{is}^{\tilde{\ell}*}$$

10.2 Higgs vertex corrections

$$i\left[\Gamma^{\tilde{\chi}_k^\pm L}_{d_i\tilde{u}_s}P_L + \Gamma^{\tilde{\chi}_k^\pm R}_{d_i\tilde{u}_s}P_R\right] \text{ with}$$

$$\Gamma^{\tilde{\chi}_k^\pm L}_{d_i\tilde{u}_s} = \sum_{j=1}^{3} V_{ji}\left(V^{\tilde{\chi}^{\pm*}}_{k2}Y^{u_j}W^{\tilde{u}*}_{j+3,s} - g_2 V^{\tilde{\chi}^{\pm*}}_{k1}W^{\tilde{u}*}_{js}\right) \quad (10.4)$$

$$\Gamma^{\tilde{\chi}_k^\pm R}_{d_i\tilde{u}_s} = U^{\tilde{\chi}^\pm}_{k2}Y^{d_i}\sum_{j=1}^{3} V_{ji}W^{\tilde{u}*}_{js}$$

$$i\left[\Gamma^{\tilde{\chi}_k^\pm L}_{u_i\tilde{d}_s}P_L + \Gamma^{\tilde{\chi}_k^\pm R}_{u_i\tilde{d}_s}P_R\right] \text{ with}$$

$$\Gamma^{\tilde{\chi}_k^\pm L}_{u_i\tilde{d}_s} = \sum_{j=1}^{3} V^*_{ji}\left(U^{\tilde{\chi}^{\pm*}}_{k2}Y^{d_j}W^{\tilde{d}*}_{j+3,s} - g_2 U^{\tilde{\chi}^{\pm*}}_{k2}W^{\tilde{d}*}_{js}\right) \quad (10.5)$$

$$\Gamma^{\tilde{\chi}_k^\pm R}_{u_i\tilde{d}_s} = V^{\tilde{\chi}^{\pm*}}_{k2}Y^{u_i}\sum_{j=1}^{3} V^*_{ji}W^{\tilde{d}*}_{js}$$

10.2. Higgs vertex corrections

In the super-CKM basis the coefficients $H^{+\,AB}_{ij}$ in (3.33) are given by

$$\begin{aligned}
H^{+\,LR}_{ij} &= \mu V^{(0)}_{ij}Y^{d_j}\cos\beta - \sum_{k=1}^{3} V^{(0)}_{jk}A^d_{ki}\sin\beta \\
H^{+\,RL}_{ij} &= \mu^* V^{(0)}_{ij}Y^{u_i*}\sin\beta - \sum_{k=1}^{3} V^{(0)}_{ki}A^{u*}_{kj}\cos\beta \\
H^{+\,LL}_{ij} &= \sin(2\beta)\frac{M_W}{\sqrt{2}g_2}V^{(0)}_{ij}\left(|Y^{u_i}|^2 + |Y^{d_j}|^2 - g_2^2\right) \\
H^{+\,RR}_{ij} &= \frac{\sqrt{2}M_W}{g_2}Y^{u_i*}V^{(0)}_{ij}Y^{d_j}
\end{aligned} \quad (10.6)$$

10.3. Loop integrals

Finally we quote our conventions for the two-point, three-point and four-point one-loop functions B_0, C_0 and D_0:

$$B_0\left(m_1^2, m_2^2\right) = 1 + \frac{m_1^2 \ln\left(\frac{Q^2}{m_1^2}\right) - m_2^2 \ln\left(\frac{Q^2}{m_2^2}\right)}{m_1^2 - m_2^2}$$

$$C_0\left(m_1^2, m_2^2, m_3^2\right) = \frac{B_0(m_1^2, m_2^2) - B_0(m_1^2, m_3^2)}{m_2^2 - m_3^2}$$

$$= \frac{m_1^2 m_2^2 \ln\left(\frac{m_1^2}{m_2^2}\right) + m_2^2 m_3^2 \ln\left(\frac{m_2^2}{m_3^2}\right) + m_3^2 m_1^2 \ln\left(\frac{m_3^2}{m_1^2}\right)}{(m_1^2 - m_2^2)(m_2^2 - m_3^2)(m_3^2 - m_1^2)}$$

$$D_0\left(m_1^2, m_2^2, m_3^2, m_4^2\right) = \frac{C_0(m_1^2, m_2^2, m_3^2) - C_0(m_1^2, m_2^2, m_4^2)}{m_3^2 - m_4^2}$$

The two-point function B_0 is UV-divergent, our definition above is $\overline{\text{MS}}$-subtracted. UV divergence and the renormalisation scale Q drop out from our results thanks to the super-GIM mechanism.

Danksagung

Ich möchte mich herzlich bei allen bedanken, die mich während meiner Zeit als Doktorand am Institut für Theoretische Teilchenphysik unterstützt haben. Insbesonder gilt mein Dank:

- Dem gesamten Institut und vor allem der Arbeitsgruppe für die gute Arbeitsatmosphäre.

- Dem Land Baden-Württemberg für die Förderung während der Doktorarbeit durch das Strukturierte Promotionskolleg (SPK).

- Meinem Bruder Stefan für das Drucken der Arbeit mit seinem Farblaserdrucker.

- Meinen Mitdiplomanden und Mitdoktoranden Lars Hofer und Dominik Scherer die mein "Schicksal" wärend den letzten vier Jahren teilten. Insbesondere gilt mein Dank Lars Hofer für die vielen hilfreichen Diskussionen sowie die unterhaltsame Begleitung auf den Konferenzen und Workshops die sich so viel kurzweiliger gestalteten.

- Ich bedanke mich bei ihm und Jennifer Girrbach für das Korrekturlesen der Doktorarbeit.

- Momchil Davidkov und Jennifer Girrbach für die gute Zusammenarbeit bei Ref. [5] und [6].

- Mikolaj Misiak für die Hilfe bei Ref. [4].

- Margarete Mühlleitner danke ich für die Übernahme des Korreferats.

- Und natürlich möchte ich mich besonders herzlich bei "meinem" Professor Ulrich Nierste für das interessante Arbeitsthema, die kontinuierliche Unterstützung, die freundliche und gute Betreuung, für viel Freiheit und Verständnis bedanken.

Bibliography

[1] A. Crivellin and U. Nierste, *Supersymmetric renormalisation of the CKM matrix and new constraints on the squark mass matrices*, Phys. Rev. **D79** (2009) 035018, [0810.1613].

[2] A. Crivellin, *CKM Elements from Squark Gluino Loops*, Proceedings of the XLIVth Rencontres der Moriond, Electroweak Interactions and Unified Theories (2009) [0905.3130].

[3] A. Crivellin and U. Nierste, *Chirally enhanced corrections to FCNC processes in the generic MSSM*, Phys. Rev. **D81** (2010) 095007, [0908.4404].

[4] A. Crivellin, *Effects of right-handed charged currents on the determinations of $|V_{ub}|$ and $|V_{cb}|$*, Phys. Rev. **D81** (2010) 031301, [0907.2461].

[5] A. Crivellin and J. Girrbach, *Constraining the MSSM sfermion mass matrices with light fermion masses*, Phys. Rev. **D81** (2010) 076001, [1002.0227].

[6] A. Crivellin and M. Davidkov, *Do squarks have to be degenerate? Constraining the mass splitting with Kaon and D mixing*, Phys. Rev. **D81** (2010) 095004, [1002.2653].

[7] F. Gabbiani and A. Masiero, *FCNC in Generalized Supersymmetric Theories*, Nucl. Phys. **B322** (1989) 235.

[8] S. Bertolini, F. Borzumati, A. Masiero, and G. Ridolfi, *Effects of supergravity induced electroweak breaking on rare B decays and mixings*, Nucl. Phys. **B353** (1991) 591–649.

[9] J. S. Hagelin, S. Kelley, and T. Tanaka, *Supersymmetric flavor changing neutral currents: Exact amplitudes and phenomenological analysis*, Nucl. Phys. **B415** (1994) 293–331.

[10] F. Gabbiani, E. Gabrielli, A. Masiero, and L. Silvestrini, *A complete analysis of FCNC and CP constraints in general SUSY extensions of the standard model*, Nucl. Phys. **B477** (1996) 321–352, [hep-ph/9604387].

[11] M. Ciuchini et. al., *Delta M(K) and epsilon(K) in SUSY at the next-to-leading order*, JHEP **10** (1998) 008, [hep-ph/9808328].

[12] M. Misiak, S. Pokorski, and J. Rosiek, *Supersymmetry and FCNC effects*, Adv. Ser. Direct. High Energy Phys. **15** (1998) 795–828, [hep-ph/9703442].

[13] F. Borzumati, C. Greub, T. Hurth, and D. Wyler, *Gluino contribution to radiative B decays: Organization of QCD corrections and leading order results*, Phys. Rev. **D62** (2000) 075005, [hep-ph/9911245].

[14] D. Becirevic et. al., $B_d - \bar{B}_d$ *mixing and the* $B_d \to J/\psi K_s$ *asymmetry in general SUSY models*, Nucl. Phys. **B634** (2002) 105–119, [hep-ph/0112303].

[15] T. Besmer, C. Greub, and T. Hurth, *Bounds on flavor violating parameters in supersymmetry*, Nucl. Phys. **B609** (2001) 359–386, [hep-ph/0105292].

[16] M. Ciuchini, E. Franco, A. Masiero, and L. Silvestrini, $b \to s$ *transitions: A new frontier for indirect SUSY searches*, Phys. Rev. **D67** (2003) 075016, [hep-ph/0212397].

[17] G. Isidori and A. Retico, $B_{s,d} \to \ell^+\ell^-$ *and* $K_L \to \ell^+\ell^-$ *in SUSY models with nonminimal sources of flavor mixing*, JHEP **09** (2002) 063, [hep-ph/0208159].

[18] A. J. Buras, T. Ewerth, S. Jager, and J. Rosiek, $K^+ \to \pi^+ \nu\bar{\nu}$ *and* $K(L) \to \pi^0 \nu\bar{\nu}$ *decays in the general MSSM*, Nucl. Phys. **B714** (2005) 103–136, [hep-ph/0408142].

[19] J. Foster, K.-i. Okumura, and L. Roszkowski, *New Higgs effects in B physics in supersymmetry with general flavour mixing*, Phys. Lett. **B609** (2005) 102–110, [hep-ph/0410323].

[20] J. Foster, K.-i. Okumura, and L. Roszkowski, *Probing the flavour structure of supersymmetry breaking with rare B-processes: A beyond leading order analysis*, JHEP **08** (2005) 094, [hep-ph/0506146].

[21] J. Foster, K.-i. Okumura, and L. Roszkowski, *Current and future limits on general flavour violation in $b \to s$ transitions in minimal supersymmetry*, JHEP **03** (2006) 044, [hep-ph/0510422].

[22] J. Foster, K.-i. Okumura, and L. Roszkowski, *New constraints on SUSY flavour mixing in light of recent measurements at the Tevatron*, Phys. Lett. **B641** (2006) 452–460, [hep-ph/0604121].

[23] M. Ciuchini and L. Silvestrini, *Upper bounds on SUSY contributions to $b \to s$ transitions from B/s - anti-B/s mixing*, Phys. Rev. Lett. **97** (2006) 021803, [hep-ph/0603114].

[24] M. Ciuchini et. al., *Next-to-leading order strong interaction corrections to the Delta(F) = 2 effective hamiltonian in the MSSM*, JHEP **09** (2006) 013, [hep-ph/0606197].

[25] M. Ciuchini et. al., *$D - \bar{D}$ mixing and new physics: General considerations and constraints on the MSSM*, Phys. Lett. **B655** (2007) 162–166, [hep-ph/0703204].

[26] W. Altmannshofer, A. J. Buras, D. M. Straub, and M. Wick, *New strategies for New Physics search in $B \to K^* \nu \bar{\nu}$, $B \to K \nu \bar{\nu}$ and $B \to X_s \nu \bar{\nu}$ decays*, JHEP **04** (2009) 022, [0902.0160].

[27] W. Altmannshofer, A. J. Buras, S. Gori, P. Paradisi, and D. M. Straub, *Anatomy and Phenomenology of FCNC and CPV Effects in SUSY Theories*, 0909.1333.

[28] L. J. Hall, V. A. Kostelecky, and S. Raby, *New Flavor Violations in Supergravity Models*, Nucl. Phys. **B267** (1986) 415.

[29] H. E. Logan and U. Nierste, *$B_{s,d} \to \ell^+ \ell^-$ in a two Higgs doublet model*, Nucl. Phys. **B586** (2000) 39–55, [hep-ph/0004139].

[30] S. Dimopoulos and H. Georgi, *Softly Broken Supersymmetry and SU(5)*, Nucl. Phys. **B193** (1981) 150.

[31] Y. Nir and N. Seiberg, *Should squarks be degenerate?*, Phys. Lett. **B309** (1993) 337–343, [hep-ph/9304307].

[32] Y. Nir and G. Raz, *Quark squark alignment revisited*, Phys. Rev. **D66** (2002) 035007, [hep-ph/0206064].

[33] E. Dudas, S. Pokorski, and C. A. Savoy, *Soft scalar masses in supergravity with horizontal $U(1)_X$ gauge symmetry*, Phys. Lett. **B369** (1996) 255–261, [hep-ph/9509410].

[34] K. Agashe and C. D. Carone, *Supersymmetric flavor models and the $B \to \Phi K(S)$ anomaly*, Phys. Rev. **D68** (2003) 035017, [hep-ph/0304229].

[35] K. Blum, Y. Grossman, Y. Nir, and G. Perez, *Combining $K - \bar{K}$ mixing and $D - \bar{D}$ mixing to constrain the flavor structure of new physics*, Phys. Rev. Lett. **102** (2009) 211802, [0903.2118].

[36] O. Gedalia, Y. Grossman, Y. Nir, and G. Perez, *Lessons from Recent Measurements of $D - \bar{D}$ Mixing*, Phys. Rev. **D80** (2009) 055024, [0906.1879].

[37] G. Senjanovic and R. N. Mohapatra, *Exact Left-Right Symmetry and Spontaneous Violation of Parity*, Phys. Rev. **D12** (1975) 1502.

[38] V. Bernard, M. Oertel, E. Passemar, and J. Stern, *Tests of non-standard electroweak couplings of right-handed quarks*, JHEP **01** (2008) 015, [0707.4194].

[39] X.-G. He, J. Tandean, and G. Valencia, *Probing New Physics in Charm Couplings with FCNC*, Phys. Rev. **D80** (2009) 035021, [0904.2301].

[40] C.-H. Chen and S.-h. Nam, *Left-right mixing on leptonic and semileptonic $b \to u$ decays*, Phys. Lett. **B666** (2008) 462–466, [0807.0896].

[41] S. Weinberg, *Implications of Dynamical Symmetry Breaking*, Phys. Rev. **D13** (1976) 974–996.

[42] E. Gildener, *Gauge Symmetry Hierarchies*, Phys. Rev. **D14** (1976) 1667.

[43] L. Susskind, *Dynamics of Spontaneous Symmetry Breaking in the Weinberg-Salam Theory*, Phys. Rev. **D20** (1979) 2619–2625.

[44] G. 't Hooft, (ed.) et. al., *Recent Developments in Gauge Theories. Proceedings, Nato Advanced Study Institute, Cargese, France, August 26 - September 8, 1979*, . New York, Usa: Plenum (1980) 438 P. (Nato Advanced Study Institutes Series: Series B, Physics, 59).

[45] S. R. Coleman and J. Mandula, *ALL POSSIBLE SYMMETRIES OF THE S MATRIX*, Phys. Rev. **159** (1967) 1251–1256.

[46] R. Haag, J. T. Lopuszanski, and M. Sohnius, *All Possible Generators of Supersymmetries of the s Matrix*, Nucl. Phys. **B88** (1975) 257.

[47] L. Hofer, U. Nierste, and D. Scherer, *Resummation of tan-beta-enhanced supersymmetric loop corrections beyond the decoupling limit*, JHEP **10** (2009) 081, [0907.5408].

[48] A. J. Buras, P. H. Chankowski, J. Rosiek, and L. Slawianowska, $\Delta M_{d,s}, B^0_{d,s} \to \mu^+\mu^-$ and $B \to X_s\gamma$ in supersymmetry at large $\tan\beta$, Nucl. Phys. **B659** (2003) 3, [hep-ph/0210145].

[49] L. J. Hall, R. Rattazzi, and U. Sarid, *The Top quark mass in supersymmetric SO(10) unification*, Phys. Rev. **D50** (1994) 7048–7065, [hep-ph/9306309].

[50] M. S. Carena, D. Garcia, U. Nierste, and C. E. M. Wagner, *Effective Lagrangian for the $\bar{t}bH^+$ interaction in the MSSM and charged Higgs phenomenology*, Nucl. Phys. **B577** (2000) 88–120, [hep-ph/9912516].

[51] W. J. Marciano and A. Sirlin, *On the Renormalization of the Charm Quartet Model*, Nucl. Phys. **B93** (1975) 303.

[52] A. Denner and T. Sack, *RENORMALIZATION OF THE QUARK MIXING MATRIX*, Nucl. Phys. **B347** (1990) 203–216.

[53] P. Gambino, P. A. Grassi, and F. Madricardo, *Fermion mixing renormalization and gauge invariance*, Phys. Lett. **B454** (1999) 98–104, [hep-ph/9811470].

[54] A. Barroso, L. Brucher, and R. Santos, *Renormalization of the Cabibbo-Kobayashi-Maskawa matrix*, Phys. Rev. **D62** (2000) 096003, [hep-ph/0004136].

[55] A. Denner, E. Kraus, and M. Roth, *Physical renormalization condition for the quark-mixing matrix*, Phys. Rev. **D70** (2004) 033002, [hep-ph/0402130].

[56] C. Hamzaoui, M. Pospelov, and M. Toharia, *Higgs-mediated FCNC in supersymmetric models with large* $\tan\beta$, Phys. Rev. **D59** (1999) 095005, [hep-ph/9807350].

[57] K. S. Babu and C. F. Kolda, *Higgs mediated $B^0 \to \mu^+\mu^-$ in minimal supersymmetry*, Phys. Rev. Lett. **84** (2000) 228–231, [hep-ph/9909476].

[58] T. Appelquist and J. Carazzone, *Infrared Singularities and Massive Fields*, Phys. Rev. **D11** (1975) 2856.

[59] U. Nierste, *Quark mixing and CP violation: The CKM matrix*, Int. J. Mod. Phys. **A21** (2006) 1724–1737, [hep-ph/0511125].

[60] G. Isidori and A. Retico, *Scalar flavor changing neutral currents in the large tan beta limit*, JHEP **11** (2001) 001, [hep-ph/0110121].

[61] P. H. Chankowski and L. Slawianowska, *$B^0_{d,s} \to \mu^-\mu^+$ decay in the MSSM*, Phys. Rev. **D63** (2001) 054012, [hep-ph/0008046].

[62] T. Banks, *Supersymmetry and the Quark Mass Matrix*, Nucl. Phys. **B303** (1988) 172.

[63] A. Masiero, P. Paradisi, and R. Petronzio, *Probing new physics through $\mu - e$ universality in $K \to l\nu$*, Phys. Rev. **D74** (2006) 011701, [hep-ph/0511289].

[64] A. Masiero, P. Paradisi, and R. Petronzio, *Anatomy and Phenomenology of the Lepton Flavor Universality in SUSY Theories*, JHEP **11** (2008) 042, [0807.4721].

[65] J. Girrbach and U. Nierste, *A critical look at $\Gamma(K \to e\nu)/\Gamma(K \to \mu\nu)$*, in preparation.

[66] F. Borzumati, G. R. Farrar, N. Polonsky, and S. D. Thomas, *Soft Yukawa couplings in supersymmetric theories*, Nucl. Phys. **B555** (1999) 53–115, [hep-ph/9902443].

[67] **Particle Data Group** Collaboration, C. Amsler et. al., *Review of particle physics*, Phys. Lett. **B667** (2008) 1.

[68] J. A. Casas, A. Lleyda, and C. Munoz, *Strong constraints on the parameter space of the MSSM from charge and color breaking minima*, Nucl. Phys. **B471** (1996) 3–58, [hep-ph/9507294].

[69] G. L. Kane, C. Kolda, and J. E. Lennon, *$B_s \to \mu^+\mu^-$ as a probe of tan(beta) at the Tevatron*, hep-ph/0310042.

[70] L. Silvestrini, *Searching for new physics in $b \to s$ hadronic penguin decays*, Ann. Rev. Nucl. Part. Sci. **57** (2007) 405–440, [0705.1624].

[71] G. L. Fogli, E. Lisi, A. Marrone, A. Palazzo, and A. M. Rotunno, *Hints of $\theta_{13} > 0$ from global neutrino data analysis*, 0806.2649.

[72] J. Girrbach, S. Mertens, U. Nierste, and S. Wiesenfeldt, *Lepton flavour violation in the MSSM*, 0910.2663.

[73] S. Dittmaier, G. Hiller, T. Plehn, and M. Spannowsky, *Charged-Higgs Collider Signals with or without Flavor*, Phys. Rev. **D77** (2008) 115001, [0708.0940].

[74] T. Plehn, M. Rauch, and M. Spannowsky, *Understanding Single Tops using Jets*, 0906.1803.

[75] M. Ciuchini et. al., *Next-to-leading order QCD corrections to Delta(F) = 2 effective Hamiltonians*, Nucl. Phys. **B523** (1998) 501–525, [hep-ph/9711402].

[76] G. Degrassi, P. Gambino, and P. Slavich, *QCD corrections to radiative B decays in the MSSM with minimal flavor violation*, Phys. Lett. **B635** (2006) 335–342, [hep-ph/0601135].

[77] M. Misiak and M. Steinhauser, *NNLO QCD corrections to the $B \to X_s \gamma$ matrix elements using interpolation in m_c*, Nucl. Phys. **B764** (2007) 62–82, [hep-ph/0609241].

[78] S. Baek, T. Goto, Y. Okada, and K.-i. Okumura, *Muon anomalous magnetic moment, lepton flavor violation, and flavor changing neutral current processes in SUSY GUT with right-handed neutrino*, Phys. Rev. **D64** (2001) 095001, [hep-ph/0104146].

[79] A. J. Buras, S. Jager, and J. Urban, *Master formulae for Delta(F) = 2 NLO-QCD factors in the standard model and beyond*, Nucl. Phys. **B605** (2001) 600–624, [hep-ph/0102316].

[80] V. Lubicz and C. Tarantino, *Flavour physics and Lattice QCD: averages of lattice inputs for the Unitarity Triangle Analysis*, Nuovo Cim. **123B** (2008) 674–688, [0807.4605].

[81] J. Virto, *Exact NLO strong interaction corrections to the Delta F=2 effective Hamiltonian in the MSSM*, JHEP **11** (2009) 055, [0907.5376].

[82] K. Anikeev et. al., *B physics at the Tevatron: Run II and beyond*, hep-ph/0201071.

[83] A. Lenz and U. Nierste, *Theoretical update of $B_s - \bar{B}_s$ mixing*, JHEP **06** (2007) 072, [hep-ph/0612167].

[84] **D0** Collaboration, V. M. Abazov et. al., *Measurement of the charge asymmetry in semileptonic B_s decays*, Phys. Rev. Lett. **98** (2007) 151801, [hep-ex/0701007].

[85] **CDF** Collaboration, A. Abulencia et. al., *Observation of $B_s - \bar{B}_s$ oscillations*, Phys. Rev. Lett. **97** (2006) 242003, [hep-ex/0609040].

[86] **CKMfitter Group** Collaboration, J. Charles et. al., *CP violation and the CKM matrix: Assessing the impact of the asymmetric B factories*, Eur. Phys. J. **C41** (2005) 1–131, [hep-ph/0406184]. 2009 update at http://ckmfitter.in2p3.fr/.

[87] M. Ciuchini et. al., *2000 CKM triangle analysis: A Critical review with updated experimental inputs and theoretical parameters*, JHEP **07** (2001) 013, [hep-ph/0012308].

[88] **CDF** Collaboration, T. Aaltonen et. al., *First Flavor-Tagged Determination of Bounds on Mixing-Induced CP Violation in $B_s^0 \to J/\psi\phi$ Decays*, Phys. Rev. Lett. **100** (2008) 161802, [0712.2397].

[89] **D0** Collaboration, V. M. Abazov et. al., *Measurement of B_s^0 mixing parameters from the flavor-tagged decay $B_s^0 \to J/\psi\phi$*, Phys. Rev. Lett. **101** (2008) 241801, [0802.2255].

[90] **UTfit** Collaboration, M. Bona et. al., *First Evidence of New Physics in $b \longleftrightarrow s$ Transitions*, PMC Phys. **A3** (2009) 6, [0803.0659].

[91] V. Tisserand, *CKM fits as of winter 2009 and sensitivity to New Physics*, 0905.1572.

[92] S. Weinberg, *Mixing angle in renormalizable theories of weak and electromagnetic interactions*, Phys. Rev. **D5** (1972) 1962–1967.

[93] S. Weinberg, *Electromagnetic and weak masses*, Phys. Rev. Lett. **29** (1972) 388–392.

[94] W. Buchmuller and D. Wyler, *CP Violation and R Invariance in Supersymmetric Models of Strong and Electroweak Interactions*, Phys. Lett. **B121** (1983) 321.

[95] J. F. Donoghue, H. P. Nilles, and D. Wyler, *Flavor Changes in Locally Supersymmetric Theories*, Phys. Lett. **B128** (1983) 55.

[96] J. Ferrandis and N. Haba, *Supersymmetry breaking as the origin of flavor*, Phys. Rev. **D70** (2004) 055003, [hep-ph/0404077].

[97] G. D'Ambrosio, G. F. Giudice, G. Isidori, and A. Strumia, *Minimal flavour violation: An effective field theory approach*, Nucl. Phys. **B645** (2002) 155–187, [hep-ph/0207036].

[98] R. S. Chivukula and H. Georgi, *Composite Technicolor Standard Model*, Phys. Lett. **B188** (1987) 99.

[99] G. D'Ambrosio, G. F. Giudice, G. Isidori, and A. Strumia, *Minimal flavour violation: An effective field theory approach*, Nucl. Phys. **B645** (2002) 155–187, [hep-ph/0207036].

[100] D. A. Demir, G. L. Kane, and T. T. Wang, *The minimal U(1)' extension of the MSSM*, Phys. Rev. **D72** (2005) 015012, [hep-ph/0503290].

[101] Y. Nir and M. P. Worah, *Probing the flavor and CP structure of supersymmetric models with $K \to \pi \nu \bar{\nu}$ decays*, Phys. Lett. **B423** (1998) 319–326, [hep-ph/9711215].

[102] A. J. Buras, G. Colangelo, G. Isidori, A. Romanino, and L. Silvestrini, *Connections between epsilon'/epsilon and rare kaon decays in supersymmetry*, Nucl. Phys. **B566** (2000) 3–32, [hep-ph/9908371].

[103] G. Colangelo, G. Isidori, and J. Portoles, *Supersymmetric contributions to direct CP violation in $K \to \pi\pi\gamma$ decays*, Phys. Lett. **B470** (1999) 134–141, [hep-ph/9908415].

[104] Y. Yamada, $b \to s\nu\bar{\nu}$ *decay in the MSSM: Implication of $b \to s\gamma$ at large $\tan \beta$*, Phys. Rev. **D77** (2008) 014025, [0709.1022].

[105] J. Prades, *Standard Model Prediction of the Muon Anomalous Magnetic Moment*, 0909.2546.

[106] S. Ferrara and E. Remiddi, *Absence of the Anomalous Magnetic Moment in a Supersymmetric Abelian Gauge Theory*, Phys. Lett. **B53** (1974) 347.

[107] J. A. Grifols and A. Mendez, *CONSTRAINTS ON SUPERSYMMETRIC PARTICLE MASSES FROM (g-2) mu*, Phys. Rev. **D26** (1982) 1809.

[108] J. R. Ellis, J. S. Hagelin, and D. V. Nanopoulos, *SPIN 0 LEPTONS AND THE ANOMALOUS MAGNETIC MOMENT OF THE MUON*, Phys. Lett. **B116** (1982) 283.

[109] R. Barbieri and L. Maiani, *The Muon Anomalous Magnetic Moment in Broken Supersymmetric Theories*, Phys. Lett. **B117** (1982) 203.

[110] D. A. Kosower, L. M. Krauss, and N. Sakai, *Low-Energy Supergravity and the Anomalous Magnetic Moment of the Muon*, Phys. Lett. **B133** (1983) 305.

[111] T. C. Yuan, R. L. Arnowitt, A. H. Chamseddine, and P. Nath, *Supersymmetric Electroweak Effects on G-2 (mu)*, Zeit. Phys. **C26** (1984) 407.

[112] M. S. Carena, G. F. Giudice, and C. E. M. Wagner, *Constraints on supersymmetric models from the muon anomalous magnetic moment*, Phys. Lett. **B390** (1997) 234–242, [hep-ph/9610233].

[113] D. Stockinger, *The muon magnetic moment and supersymmetry*, J. Phys. **G34** (2007) R45–R92, [hep-ph/0609168].

[114] S. Marchetti, S. Mertens, U. Nierste, and D. Stockinger, *Tan(beta)-enhanced supersymmetric corrections to the anomalous magnetic moment of the muon*, Phys. Rev. **D79** (2009) 013010, [0808.1530].

[115] A. Dedes, H. K. Dreiner, and U. Nierste, *Correlation of $B_s \to \mu^+\mu^-$ and $(g-2)(\mu)$ in minimal supergravity*, Phys. Rev. Lett. **87** (2001) 251804, [hep-ph/0108037].

[116] D. Eriksson, F. Mahmoudi, and O. Stal, *Charged Higgs bosons in Minimal Supersymmetry: Updated constraints and experimental prospects*, JHEP **11** (2008) 035, [0808.3551].

[117] T. Aoyama, M. Hayakawa, T. Kinoshita, and M. Nio, *Revised value of the eighth-order QED contribution to the anomalous magnetic moment of the electron*, Phys. Rev. **D77** (2008) 053012, [0712.2607].

[118] M. Cadoret et. al., *Combination of Bloch oscillations with a Ramsey-Bordé interferometer : new determination of the fine structure constant*, Phys. Rev. Lett. **101** (2008) 230801, [0810.3152].

[119] J. A. Aguilar-Saavedra et. al., *Supersymmetry parameter analysis: SPA convention and project*, Eur. Phys. J. **C46** (2006) 43–60, [hep-ph/0511344].

[120] A. De Roeck et. al., *Supersymmetric benchmarks with non-universal scalar masses or gravitino dark matter*, Eur. Phys. J. **C49** (2007) 1041–1066, [hep-ph/0508198].

[121] T. Hurth and W. Porod, *Flavour violating squark and gluino decays*, JHEP **08** (2009) 087, [0904.4574].

[122] S. Khalil and O. Lebedev, *Chargino Contributions to Epsilon and Epsilon-Prime*, Phys. Lett. **B515** (2001) 387–394, [hep-ph/0106023].

[123] **BABAR** Collaboration, B. Aubert et. al., *Evidence for D^0-$\bar{D}^0$ Mixing*, Phys. Rev. Lett. **98** (2007) 211802, [hep-ex/0703020].

[124] **Belle** Collaboration, M. Staric et. al., *Evidence for D^0 - $\bar{D}^0$ Mixing*, Phys. Rev. Lett. **98** (2007) 211803, [hep-ex/0703036].

[125] **BELLE** Collaboration, K. Abe et. al., *Measurement of D^0 - $\bar{D}^0$ mixing in $D^0 \to K_s \pi^+ \pi^-$ decays*, Phys. Rev. Lett. **99** (2007) 131803, [0704.1000].

[126] **Heavy Flavor Averaging Group** Collaboration, E. Barberio et. al., *Averages of $b-$hadron and $c-$hadron Properties at the End of 2007*, 0808.1297, and online update at http://www.slac.stanford.edu/xorg/hfag.

[127] O. Gedalia, L. Mannelli, and G. Perez, *Covariant Description of Flavor Violation at the LHC*, 1002.0778.

[128] W. Altmannshofer, A. J. Buras, and D. Guadagnoli, *The MFV limit of the MSSM for low tan(beta) : Meson mixings revisited*, JHEP **11** (2007) 065, [hep-ph/0703200].

[129] B. Grzadkowski and M. Misiak, *Anomalous Wtb coupling effects in the weak radiative B- meson decay*, Phys. rev. **D78** (2008) 077501, [0802.1413].

[130] W. Buchmuller and D. Wyler, *Effective Lagrangian Analysis of New Interactions and Flavor Conservation*, Nucl. Phys. **B268** (1986) 621.

[131] B. Dassinger, R. Feger, and T. Mannel, *Complete Michel Parameter Analysis of inclusive semileptonic $b \to c$ transition*, 0803.3561.

[132] M. C. Arnesen, B. Grinstein, I. Z. Rothstein, and I. W. Stewart, *A precision model independent determination of $|V_{ub}|$ from $B \to \pi e \nu$*, Phys. Rev. Lett. **95** (2005) 071802, [hep-ph/0504209].

I want morebooks!

Buy your books fast and straightforward online - at one of world's fastest growing online book stores! Environmentally sound due to Print-on-Demand technologies.

Buy your books online at
www.morebooks.shop

Kaufen Sie Ihre Bücher schnell und unkompliziert online – auf einer der am schnellsten wachsenden Buchhandelsplattformen weltweit! Dank Print-On-Demand umwelt- und ressourcenschonend produziert.

Bücher schneller online kaufen
www.morebooks.shop

KS OmniScriptum Publishing
Brivibas gatve 197
LV-1039 Riga, Latvia
Telefax +371 686 204 55

info@omniscriptum.com
www.omniscriptum.com

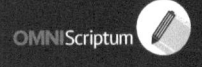

Printed by Books on Demand GmbH, Norderstedt / Germany